DRUG METABOLISM

DRUG METABOLISM

Chemical and Enzymatic Aspects

TEXTBOOK EDITION

Jack P. Uetrecht
University of Toronto
Ontario, Canada

William Trager
University of Washington
Seattle, Washington, USA

informa
healthcare

New York   London

Informa Healthcare USA, Inc.
52 Vanderbilt Avenue
New York, NY 10017

© 2007 by Informa Healthcare USA, Inc.
Informa Healthcare is an Informa business

No claim to original U.S. Government works
Printed in the United States of America on acid-free paper
10 9 8 7 6 5 4 3 2 1

International Standard Book Number-10: 1-4200-6103-8
International Standard Book Number-13: 978-1-4200-6103-1

Library of Congress Cataloging-in-Publication Data

Uetrecht, Jack P.
Drug metabolism : chemical and enzymatic aspects / edited [i.e. compiled] Jack P. Uetrecht, William Trager. – Textbook ed.
    p. cm.
  Includes bibliographical references and index.
  ISBN-13: 978-1-4200-6103-1 (hb : alk. paper)
  ISBN-10: 1-4200-6103-8 (hb : alk. paper)
  1. Drugs – Metabolism – Textbooks.   I. Trager, William, 1937–   II. Title.

RM301.55.U382 2007
615'.7 – dc22   2007016201

Visit the Informa Web site at
www.informa.com

and the Informa Healthcare Web site at
www.informahealthcare.com

Preface

Drugs and other xenobiotics can exert a wide variety of pharmacological and toxic effects. In order to understand these effects it is necessary to understand both the structural parameters that are a direct cause of these effects as well as the factors that control the concentration of the drug in the body, such as absorption, metabolism, and elimination. To fully appreciate drug action requires at least minimal expertise in a variety of disciplines such as chemistry, biochemistry, kinetics and biology. While complicated and demanding in its breadth such knowledge is central to the knowledge base of advanced students of Pharmacy, Pharmacokinetics, Medicinal Chemistry, Pharmacology and Toxicology. However, texts that focus on covering these disciplines in rationalizing drug action are rare. Furthermore, many of the effects are due to metabolites rather than due to the parent drug/xenobiotic and some metabolites, or intermediates that lead to metabolites, are chemically reactive. Therefore, when considering the effects of an agent, all of the metabolites that the body produces from the agent must also be taken into consideration.

To try and address the gap in multidisciplinary knowledge required, as the title suggests, the focus of this book is on the chemistry, enzymology and to a lesser extent the kinetics of drug metabolism. As indicated above an understanding of this subject at a minimum requires a basic understanding of the chemistry involved. It is also important that these processes be placed into a biological context. Therefore, Chapter 2 entitled "Background for Nonchemists" and Chapter 3 entitled "Background for Chemists" attempt to explicitly confront these issues and provide the necessary background and context. Since chemically reactive metabolites have major implications for toxicity and since understanding their generation and properties requires the spectrum of disciplines outlined above a chapter is devoted to reactive metabolites. Finally, the only way to master the subject is with practice. Sample problems with answers are provided to facilitate this process.

Jack P. Uetrecht
William Trager

Contents

1
Introduction

The response of different patients to a drug varies widely and, depending on the drug category, from 20% to 75% of patients do not have a therapeutic response (1). In addition, many patients will have an adverse reaction to a drug. There are many reasons for these interindividual differences in drug response, both pharmacokinetic and pharmacodynamic. In order to begin to understand these differences, it is essential to understand what happens to a drug in the body. Most drugs are given orally and some drugs have variable and incomplete absorption. The major determinant of absorption is the physical properties of the drug. Once in the body, most drugs are converted to multiple metabolites. This process can begin in the intestine or liver before the drug even enters the blood stream. Metabolism is often required in order for the body to eliminate a drug. However, some drugs are prodrugs, i.e., in order to exert a therapeutic effect they require metabolism to convert them to an active agent. Examples include enalapril, which has better oral bioavailability than the active agent, enalaprilate, and is readily activated by hydrolysis, and codeine, which must be metabolized to morphine in order to be an effective analgesic (Fig. 1.1).

The enzyme that converts codeine to morphine is polymorphic, and about 7% of the North American population lacks the cytochrome P450 (CYP2D6) needed to perform this conversion; now we understand why codeine does not work in these patients while others have an exaggerated response because of very high levels of CYP2D6 (2). There are several such metabolic enzymes that have common genetic polymorphisms caused by differences in a single nucleotide that influences enzyme expression or protein structure. Classically, this leads to a bi- or even trimodal distribution of enzyme activity in a population. In principle, this could be due to differences in intrinsic activity of the enzyme, but in most cases the genetic variant has very low levels of the enzyme, usually because of rapid protein degradation (3). Such polymorphisms can result in an interindividual difference of more than 100 fold in the blood levels of a drug that is metabolized by a polymorphic enzyme. Examples of polymorphic metabolic enzymes are listed in Table 1.1 (2,4,5). Other metabolic enzymes have more of a Gaussian distribution of enzyme activity in a population, i.e., they do not exhibit a classic bimodal distribution. This is because there are no common variants leading to large differences in enzyme activity and/or because the expression of the

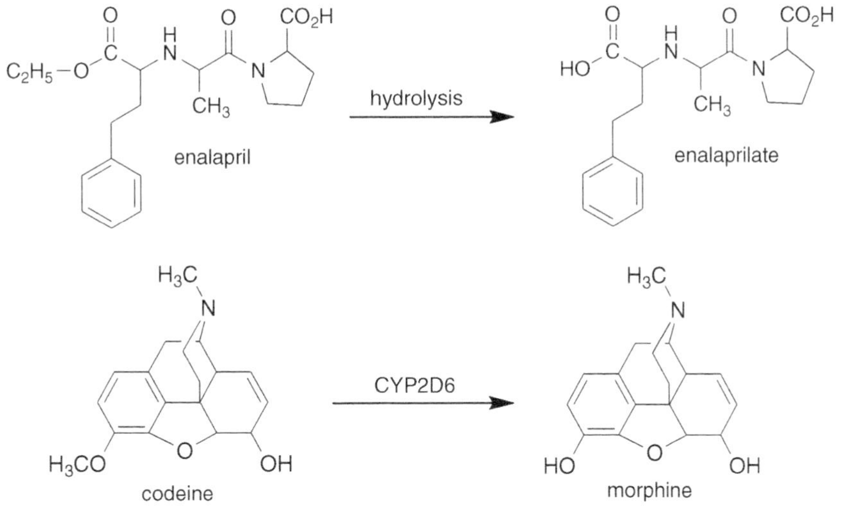

FIGURE 1.1   Metabolic conversion of prodrugs to pharmacologically active agents.

enzyme is strongly influenced by environmental factors. A good example of such an enzyme is CYP3A4 whose activity also varies greatly from one individual to another. In a recent study, it was found that the CYP3A4 activity of microsomes from different human livers varied by more than a factor of 100 and this correlated strongly with the level of CYP3A4 protein; however, no single factor was found to be responsible for this large variation (6).

Furthermore, some metabolic pathways, such as glucuronidation and amino acid conjugation, are deficient at birth thereby making newborns more sensitive to drugs that are cleared by the enzymes involved. In the case of glucuronidation and newborns, this is particularly important because glucuronidation is the primary mechanism for the elimination of bilirubin, the breakdown product of hemoglobin, and the increase in the levels of bilirubin leads to jaundice. Another aspect of drug variability that impinges upon therapeutics is that

TABLE 1.1   Examples of Common Polymorphic Metabolic Enzymes

Enzyme	Effect of impaired metabolizer phenotype
CYP 2A6	Decreased cigarette consumption in smokers, easier to stop smoking
CYP 2C9	Exaggerated response to warfarin and phenytoin
CYP 2C19	Increased efficacy of omeprazole, increased toxicity of mephenytoin
CYP 2D6	Absence of codeine efficacy, no effect of encainide, increased levels of tricyclic antidepressants, fluoxetine, phenothiazines
Pseudocholinesterase	Sustained paralysis to succinylcholine, possible increased toxicity of cocaine
Epoxide hydrolase	Unknown
UDP-glucuronosyl-transferase 1A1	Increased toxicity of irinotecan, increased levels of bilirubin
N-acetyltransferase 2	Increased risk of hydralazine-induced lupus, increased levels of isoniazid with an increased risk of neurotoxicity, increased risk of bladder cancer in individuals exposed to aromatic amines
Thiopurine methyltransferase	Increased risk of serious toxicity to mercaptopurine and azathioprine

many drugs and other xenobiotics, i.e., chemicals that are foreign to the body, can increase or decrease the levels/activity of metabolic enzymes leading to drug–drug interactions.

Although many drugs require metabolism in order for them to be cleared from the body, many adverse effects of drugs are mediated by metabolites. In particular, it appears that most idiosyncratic drug reactions are caused by chemically reactive metabolites of drugs, and interindividual differences in metabolism of a drug may contribute to the idiosyncratic nature of these adverse reactions. However, the mechanisms of idiosyncratic reactions are not really understood, and even though there is evidence that reactive metabolites are involved, genetic polymorphisms in enzymes responsible for forming the reactive metabolite that appears to be responsible for a given idiosyncratic reaction do not appear to be the major factor leading to the idiosyncratic nature of these adverse reactions (7).

The goal of this book is to provide a basic understanding of how drugs are converted to metabolites and how these transformations can change the physical and pharmacological properties of the drug. In order to gain a basic understanding of drug metabolism, it is essential to have a basic understanding of chemistry. To make this material more accessible to nonchemists, in Chapter 2 we review the basic chemical principles required to fully appreciate the subsequent chapters. It is also important to be able to mathematically describe the rates of metabolism and other processes that control the concentration of a drug and therefore a review of pharmacokinetics is provided in Chapter 3. Drug metabolism is commonly divided into phase I (oxidation, reduction, and hydrolysis) and phase II (conjugation); this implies that phase I metabolism occurs before phase II metabolism. However, if a drug can undergo phase II metabolism, phase I metabolism may make a minor contribution to the clearance of a drug and we have chosen to organize the pathways according to their chemical nature, i.e., oxidation, etc. As mentioned above, many adverse effects of drugs appear to be due to chemically reactive metabolites; therefore, the last chapter is a discussion of reactive metabolites.

Although the emphasis is on drugs, the same principles apply to any xenobiotic agent. The major difference between drugs and other xenobiotics such as environmental toxins is the dose. The dose of common drugs is usually on the order of 100 mg/day and can be more than a gram a day; in contrast, exposure to most other xenobiotics is typically much lower.

REFERENCES

1. Spear BB, Heath-Chiozzi M, Huff J. Clinical application of pharmacogenetics. Trends Mol Med 2001;7(5):201–204.
2. Wilkinson GR. Drug metabolism and variability among patients in drug response. N Engl J Med 2005; 352(21):2211–2221.
3. Weinshilboum R, Wang L. Pharmacogenetics: Inherited variation in amino acid sequence and altered protein quantity. Clin Pharmacol Ther 2004; 75(4):253–258.
4. Evans WE, Johnson JA. Pharmacogenomics: The inherited basis for interindividual differences in drug response. Annu Rev Genomics Hum Genet 2001; 2:9–39.
5. Weinshilboum R, Wang L. Pharmacogenomics: Bench to bedside. Nat Rev Drug Discov 2004; 3(9):739–748.
6. He P, Court MH, Greenblatt DJ, et al. Factors influencing midazolam hydroxylation activity in human liver microsomes. Drug Metab Dispos 2006; 34(7):1198–1207.
7. Uetrecht J. Idiosyncratic drug reactions: Current understanding. Annu Rev Pharmacol Toxicol 2007; 47:513–539.

2

Background for Nonchemists

Although nonchemists are usually intimidated by chemical structures, drugs are chemicals and a basic understanding of chemistry is necessary to understand the properties of drugs. The name of a drug is arbitrary and provides little or no information about the drug, but the structure defines a drug and provides many clues as to the likely properties of the drug. It is a valuable skill to be able to look at the structure of a drug and be able to predict with a reasonable degree of certainty many of the properties of that drug; one of the goals of this book is to facilitate developing this skill and to make it more accessible to nonchemists. Although the focus of this book is on drugs, in principle, drugs are no different than other xenobiotics (any foreign compound, be it from an herbal product or a chemical waste).

USING THE STRUCTURE OF A DRUG TO PREDICT PROPERTIES SUCH AS WATER SOLUBILITY

The properties and pharmacological effects of drugs are due to their interactions with other molecules such as enzymes and receptors. Water solubility, which is a very important property of a drug, is based on interactions with water. Ion–dipole interactions are the strongest and therefore drugs that are mostly charged at physiological pH usually have high water solubility. The next strongest interaction is hydrogen bonds. Therefore, drugs that contain O–H or N–H groups are, in general, more water soluble than those that do not have such groups. However, the presence of one O–H group on a large molecule is not sufficient to confer high water solubility. Solubility is based on the balance of bonds formed and broken. Strong interactions with water molecules obviously promote water solubility, but interactions between water molecules must be broken to form these new interactions. Even hydrophobic drugs interact with water, but the energy of the bonds between water molecules that must be broken in the process of dissolving the hydrophobic drug in water is much greater than that of the bonds between the drug and water; therefore, hydrophobic drugs have very low water solubility. As shown in Figure 2.1, 1-butanol is surrounded by water molecules. There are hydrogen bonds between the water molecules and between

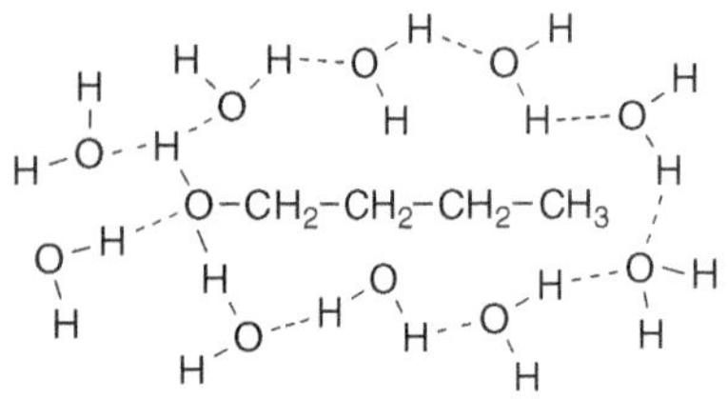

FIGURE 2.1 1-Butanol surrounded by water molecules showing possible hydrogen bond interactions.

water molecules and the OH of butanol, but the interactions between the water molecules and the hydrocarbon part of butanol are weak and interactions between water molecules have to be broken in order to dissolve the butanol. 1-Butanol is soluble in water, but unlike ethanol, it is not miscible in all proportions in water and alcohols with a longer alkyl chain are less soluble.

Likewise, another factor that affects water solubility is the strength of interactions between the drug molecules that must be broken when the drug dissolves. One clue to the strength of these interactions is the melting point of a drug. Other factors being equal, liquids are more water soluble than solids and, with the exception of salts, drugs with a high melting point usually have low water solubility. For example, in a comparison of sulfadiazine and sulfamethazine (Fig. 2.2), the only difference between the two molecules is the presence two methyl groups on sulfamethazine. Even though the methyl groups are "hydrophobic" and decrease the strength of the molecule's interaction with water, the methyl groups also decrease the strength of interactions between the sulfa molecules in the solid as evidenced by the lower melting point of sulfamethazine (176° C for sulfamethazine vs 252–256° C for sulfadiazine) and the net effect is that sulfamethazine is more soluble than sulfadiazine (2.4 mg/L for sulfamethazine vs. 0.5 mg/L for sulfadiazine). In summary, solubility is based on the net change in energy of the process, and in this case the decrease in the strength of interactions of the drug with water caused by the introduction of methyl groups is more than compensated for by the decrease in the strength of the interactions that need to be broken as the solid dissolves in water.

In larger molecules, hydrogen bonds or even ion–dipole interactions can occur within the molecule such that they interact less with water or other molecules. For example, the 6-glucuronide conjugate of morphine is much more lipophilic than would be expected from its structure (it has one negative charge, one positive charge and several hydroxy groups) and this lipophilicity allows it to readily pass the blood-brain barrier. This is presumably because the molecule can fold up on itself in such a way that the negative and positive charges interact thus decreasing their interaction water. In addition, it is likely there are also several other intramolecular hydrogen bonds that decrease the interactions with water. Although van der Waals forces are weaker, they can play an important role.

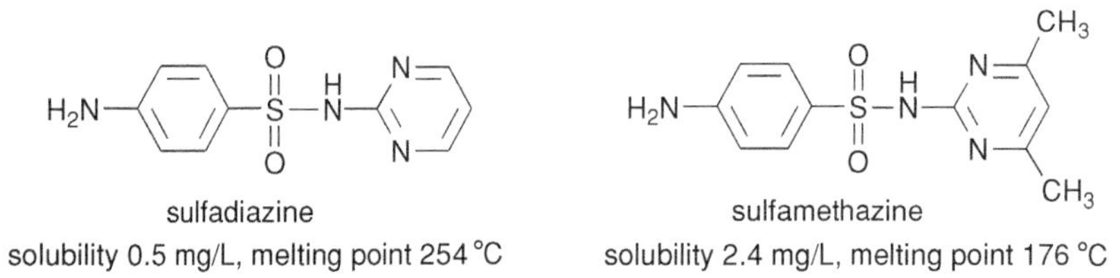

sulfadiazine
solubility 0.5 mg/L, melting point 254 °C

sulfamethazine
solubility 2.4 mg/L, melting point 176 °C

FIGURE 2.2 Relationship between solubility and melting point as demonstrated by the comparison of sulfadiazine and sulfamethazine.

van der Waals forces are due to the polarizability of the electron cloud and are therefore more important for larger atoms because the outer electrons are further from the nucleus and therefore less tightly bound. Although such interactions are less important for water solubility, they can have quite significant effects on other types of interactions. For example, carbon tetrachloride is a much better solvent for fats than hexane, presumably because chlorine atoms are much larger and their electron cloud is much more polarizable than that of hydrogen atoms. Furthermore, the addition of chlorine to a molecule markedly increases its retention time on reverse phase HPLC columns where the interaction is between the drug and an alkyl chain attached to a solid support. Likewise, the presence of a chlorine atom in a specific location can markedly increase the interaction of a drug with a receptor.

PREDICTION OF CHARGE FROM pK_a

As indicated above, charge has a major effect on the properties of a drug. The presence of a charge not only increases water solubility, but also has a major effect on its absorption, distribution, excretion, etc., and many metabolic pathways introduce a charge onto a drug in order to increase excretion, especially renal excretion. It is therefore important to be able to look at the structure of a drug and be able to predict whether it will be mostly charged at physiological pH. Although some drugs have a permanent charge (e.g., a quaternary ammonium salt, see next section), in most cases, if a drug is charged it is because it is either an acid or a base.

    An acid is any chemical entity that can donate a proton and a base is any chemical entity that can accept a proton. The strength of an acid is designated by its pK_a, which is defined by the following relationship:

$$p K_a = pH + \log \text{(concentration of protonated form/concentration of unprotonated form)}$$

From this relationship, it is clear that both protonated (acidic) and unprotonated (basic) forms exist in equilibrium; thus, every acid has a conjugate base (the unprotonated form) and every base has a conjugate acid (the protonated form, Fig. 2.3).

    The conjugate acid of a weak base is a strong acid and the conjugate acid of a strong base is a weak acid. What is important for this discussion is whether a molecule is mostly charged at physiological pH, which for blood is tightly controlled at pH 7.4. The pK_a of a drug is the decisive property for making this judgment. As indicated in the relationship presented above, when pK_a equals pH, equal concentrations of protonated and unprotonated drug will be present (log 1 = 0). Because this is a log scale, at pH 7.4 if an acid has a pK_a of 5.4 then the unprotonated form of the drug will predominate by a ratio of 100:1, whereas if the pK_a is 9.4 then the protonated form of the drug will predominate by a ratio of 100:1. The trick then is to decide whether it is the protonated or the unprotonated form that is charged. Clearly the protonated form, or conjugate acid, will always bear one more

A–H $\rightleftharpoons$ A$^-$ + H$^+$
acid   conjugate base

B + H$^+$ $\rightleftharpoons$ B$^+$H
base   conjugate acid

FIGURE 2.3   For every acid there is a conjugate base and for every base there is a conjugate acid; the difference between an acid and a base is which form is charged.

positive charge than the unprotonated, or conjugate base, form. But that does not mean that the protonated form is charged. For example, acetic acid is the protonated form of the conjugate acid–conjugate base pair. It bears one more proton than the unprotonated conjugate base form, acetate anion. But it is neutral and this is generally true for what we commonly recognize as organic acids. Conversely, aniline is the unprotonated conjugate base form of the protonated aniline–aniline, conjugate acid–conjugate base pair, but unlike acetate anion it is neutral, while the conjugate acid form, protonated aniline, is charged. In general, the conjugate acids of nitrogen-containing drugs are positively charged. Now if we consider the pK_a values of acetic acid and aniline (protonated aniline), we can arrive at definitive conclusions regarding whether or not they are charged at physiological pH. Acetic acid has a pK_a of approximately 5, which means that it will virtually all be in the unprotonated charged form of acetate anion at pH 7.4. Protonated aniline also has a pK_a of approximately 5, which in this case means that at pH 7.4, virtually all of it will be present in the neutral unprotonated conjugate base form, aniline. If you look up the pK_a of a drug in a reference book, the pK_a value will always refer to the protonated, or conjugate acid, form of that drug. However, especially in the case of amines, the reference may not specify the structure of the protonated form of the base that the pK_a measures. Therefore, it will be up to you to decide whether the pK_a is for an acid or the conjugate acid of a base, i.e., A–H or B^+H (Fig. 2.3). Only then can you determine the degree to which it will be ionized at pH 7.4. Regarding your decision, it is well to remember that in the vast majority of cases it is almost as difficult to remove a proton from a neutral nitrogen atom as it is to remove a proton from a carbon atom. There are cases, however, where a N–H bond of a neutral nitrogen-containing compound or even a C–H bond of a carbon-containing compound has been sufficiently weakened such that it will dissociate into a proton and the residual anion in water. These are invariably compounds having structural features that are highly efficient in stabilizing the anionic conjugate base by resonance. Several examples will be presented later in the text.

Similar to the analysis above, phenol and the conjugate acid of amphetamine both have a pK_a of 10 (Fig. 2.6), but phenol is un-ionized at pH 7.4 and amphetamine is mostly ionized. When the pK_a of a base is reported, as in Figure 2.6, it is understood that it represents the pK_a of the protonated form and is calculated from the ratio of protonated to unprotonated as shown in the equation above.

Virtually all drugs that are bases (of the type B in Fig. 2.3) are so because they contain nitrogen, but not all nitrogen-containing drugs are mostly ionized at pH 7.4. Aliphatic amines are strong bases because the nitrogen has a lone pair of electrons that can be used to form a bond with a hydrogen ion (proton) as shown in Figure 2.4 and therefore they are mostly ionized. An aromatic amine has the nitrogen attached to an aromatic ring. The lone pair of electrons is delocalized into the aromatic ring, which makes them less available to form a bond with a hydrogen ion and therefore aromatic amines are weak bases and mostly un-ionized at pH 7.4. Electron-withdrawing groups, such as a nitro group, further decrease the electron density on the nitrogen and further weaken the basicity of the amine. An amide has the nitrogen next to a carbonyl group. The lone pair of electrons is delocalized into the carbonyl group (Fig. 2.4) to the extent that amides are essentially not basic at all in a biological system.

In short, organic bases virtually always contain a nitrogen atom, and organic nitrogen-containing compounds are mostly ionized at pH 7.4 unless there is some electron-withdrawing group or conjugation with an aromatic ring or double bond that reduces the electron density on the nitrogen. It does not matter if the nitrogen is part of a ring as long as there are no double bonds conjugated with the nitrogen, that can delocalize the lone pair of electrons. The amines shown in Figure 2.4 are primary amines; however, the same principle applies if the amine is secondary (two carbons attached to the nitrogen) or tertiary aliphatic amine $R - \overset{|}{\underset{|}{C}} - \overset{..}{N}H_2 \xrightarrow{\ H^+\ } R - \overset{|}{\underset{|}{C}} - \overset{+}{N}H_3$ pK$_a$ ~10 (strong base)

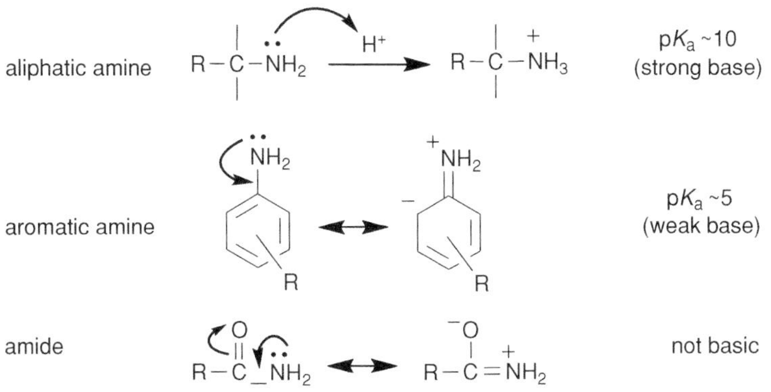

aromatic amine   pK$_a$ ~5 (weak base)

amide $R - \overset{O}{\underset{||}{C}} - \overset{..}{N}H_2 \longleftrightarrow R - \overset{\overset{-}{O}}{\underset{||}{C}} = \overset{+}{N}H_2$ not basic

FIGURE 2.4 It is the density of the lone pair of electrons on nitrogen that determines the basic character of amines.

(three carbons attached to the nitrogen). On the other hand, in a quaternary ammonium salt in which there are four carbons attached to the nitrogen, the nitrogen lone pair of electrons is used to form the fourth bond to carbon and thus is not available to bond a proton. Therefore quaternary ammonium salts are positively charged and are not basic.

For something to be an acid (of the type A–H in Fig. 2.3), it has to lose a positively charged proton and stabilize the negative charge that is left behind. Most organic acids have hydrogen attached to an oxygen atom because oxygen is highly electronegative and can more readily accept a negative charge than most elements. The simplest such structure is an alcohol (Fig. 2.5); however, the negative charge left on the oxygen after an alcohol loses a proton cannot be stabilized by delocalization and so alcohols do not have appreciable acidity in a biological system. In a phenol, the OH is attached to an aromatic ring and now the negative charge remaining on the oxygen atom after the loss of a proton can be delocalized into the aromatic ring. This makes phenol a weak acid. If there are strong electron-withdrawing groups on the benzene ring of a phenol, it can be a strong acid. When the OH is next to a carbonyl group, it is called a carboxylic acid. As the name implies, carboxylic acids are acids; this is because the negative charge is equally shared by two oxygen atoms. *Therefore, phenols are stronger acids than alcohols for the same reason that aromatic amines are weaker bases than aliphatic amines, and amides are not basic for the same reason that carboxylic acids are strong acids.*

By understanding the basis for the basicity and acidity of these six functional groups, it is possible to predict from the structure of most drugs as to whether they are mostly

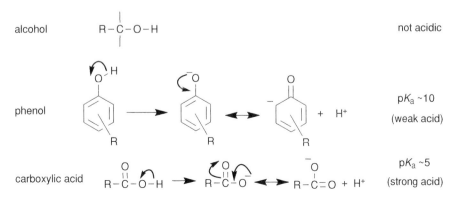

alcohol $R - \overset{|}{\underset{|}{C}} - O - H$ not acidic

phenol   pK$_a$ ~10 (weak acid)

carboxylic acid $R - \overset{O}{\underset{||}{C}} - O - H \longrightarrow R - \overset{O}{\underset{||}{C}} - O \longleftrightarrow R - \overset{\overset{-}{O}}{\underset{||}{C}} = O + H^+$ pK$_a$ ~5 (strong acid)

FIGURE 2.5 It is the density of the lone pair of electrons on nitrogen that determines the basic character of amines.

ionized at pH 7.4; however, these six functional groups do not cover all possibilities and additional examples are useful. Figure 2.6 shows several organic acids and bases. Although phenol itself is a weak acid, picric acid with its three nitro groups, which can delocalize the negative charge, is a very strong acid. Likewise, trifluoroacetic acid, a metabolite of halothane, is a very strong acid because of the electron-withdrawing effect of the three fluorine atoms. A few drugs, such as aztreonam, are sulfonic acids, which are also very strong acids (aztreonam also has an acidic carboxylic acid and together they make the oral bioavailability of azetreonam quite low). Although a benzene ring significantly increases the acidity of a phenol over an alcohol because it allows the negative charge to be delocalized, it does not significantly increase the acidity of a carboxylic acid because it does not help to delocalize the negative charge (Fig. 2.6), and the major reason that benzoic acid is slightly more acidic than acetic acid is because the sp^2 carbon of the benzene ring is more electronegative than the sp^3 carbon of the methyl group. In most drugs that are acids, the acidic proton is bound to an oxygen, the electronegativity of which helps to accommodate the charge; however, organic acids can have a hydrogen attached to a less-electronegative atom such as nitrogen, sulfur, or even carbon if the negative charge remaining after the loss of the proton is sufficiently stabilized. For example, in phenylbutazone the acidic hydrogen is attached to a carbon and the negative charge is stabilized by the flanking carbonyl groups. In sulfamethoxazole, the acidic hydrogen is attached to nitrogen and the negative charge is stabilized by the electron-withdrawing effect of the SO_2 group as well as delocalization into the oxazole ring. Phenobarbital is also acidic analogous to phenylbutazone. Thiols are also weakly acidic. Although not as electronegative as oxygen, the strength of the S–H bond is much weaker than that of the O–H bond. Thiophenol is more acidic than cysteine for the same reason that phenol is more acidic than an alcohol; however, the effect is not as great because the orbitals of the sulfur are much further from the nucleus and therefore do not overlap with those of the aromatic ring. Although three nitro groups made picric acid a very strong acid, one nitro group has much less of an effect.

Unlike acids where the acidic proton can be bound to several different atoms, virtually all drugs that are bases are basic because of a nitrogen atom. As expected from the previous discussion, amphetamine is a strong base whose conjugate acid has a pK_a of 10. The guanidine group of arginine is even more basic. This is because when the –NH group is protonated, all three of the nitrogen atoms attached to the carbon can equally share the positive charge and even the carbon atom itself bears some of the charge. The bicyclic ring system of cocaine does not significantly affect the basicity of the bridgehead nitrogen and cocaine is a strong base. Although clonidine has a guanidine group similar to arginine, the aromatic ring decreases the pK_a to 8.3. The pyridine ring (six-membered ring) of nicotine is a weak base similar to aniline because the lone pair of electrons is delocalized, while the pyrrolidine nitrogen (saturated five-membered ring) is a strong base. The aromatic amine in ketoconazole has a pK_a of 2.9, which is too low to be highly ionized even in the low pH of the stomach, but ketoconazole also contains an imidazole ring. Most heterocyclic rings with a double bond to the nitrogen are weak bases, but imidazole contains two nitrogens, and when one is protonated, it can be resonance stabilized by the second (similar to guanidine). Imidazole itself has a pK_a of 6.95, and the imidazole ring in ketoconazole, with a pK_a of 6.5, is critical for the bioavailability of the drug. Ketoconazole is mostly lipophilic with low water solubility, but in the acidic pH of the stomach, the imidazole ring is ionized and this allows it to dissolve. In patients who take other medications to decrease stomach acidity, ketoconazole does not dissolve and thus its bioavailability is very low. The imine in diazepam is conjugated with the aromatic ring and its pK_a is only 3.4, while the other nitrogen, which is an amide, is not basic in a biological system.

The preceding discussion is based on an individual acidic or basic group, but molecules can posses more than one such group, e.g., simple amino acids (Fig. 2.7). It

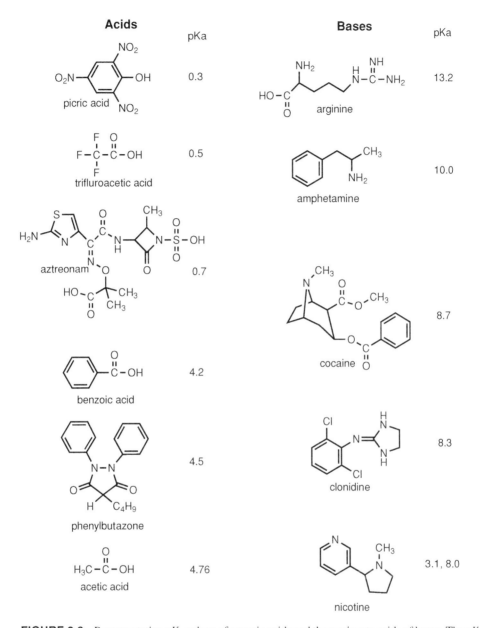

FIGURE 2.6 Representative pK_a values of organic acids and the conjugate acids of bases. The pK_a values of organic acids and bases were obtained from Perrin et al. (1), of drugs from Foye et al. (2), and of amino acids from Merck Index (3).

would appear that if a molecule has both an acidic and a basic functional group, they would neutralize each other. Indeed this is the case, e.g., alanine. Such molecules are called amphoteric because they are both acids and bases and zwitterions when they have both a positive and negative charge. If alanine is dissolved in a strongly acidic solution, pH < 1, both the amino group and the acidic group will be protonated. If the solution is then titrated with base, two pK_as will be found: one for the ionization of the carboxyl group, at ~2.4, and the other for the ionization of the protonated amino group, at ~9.6. The average of

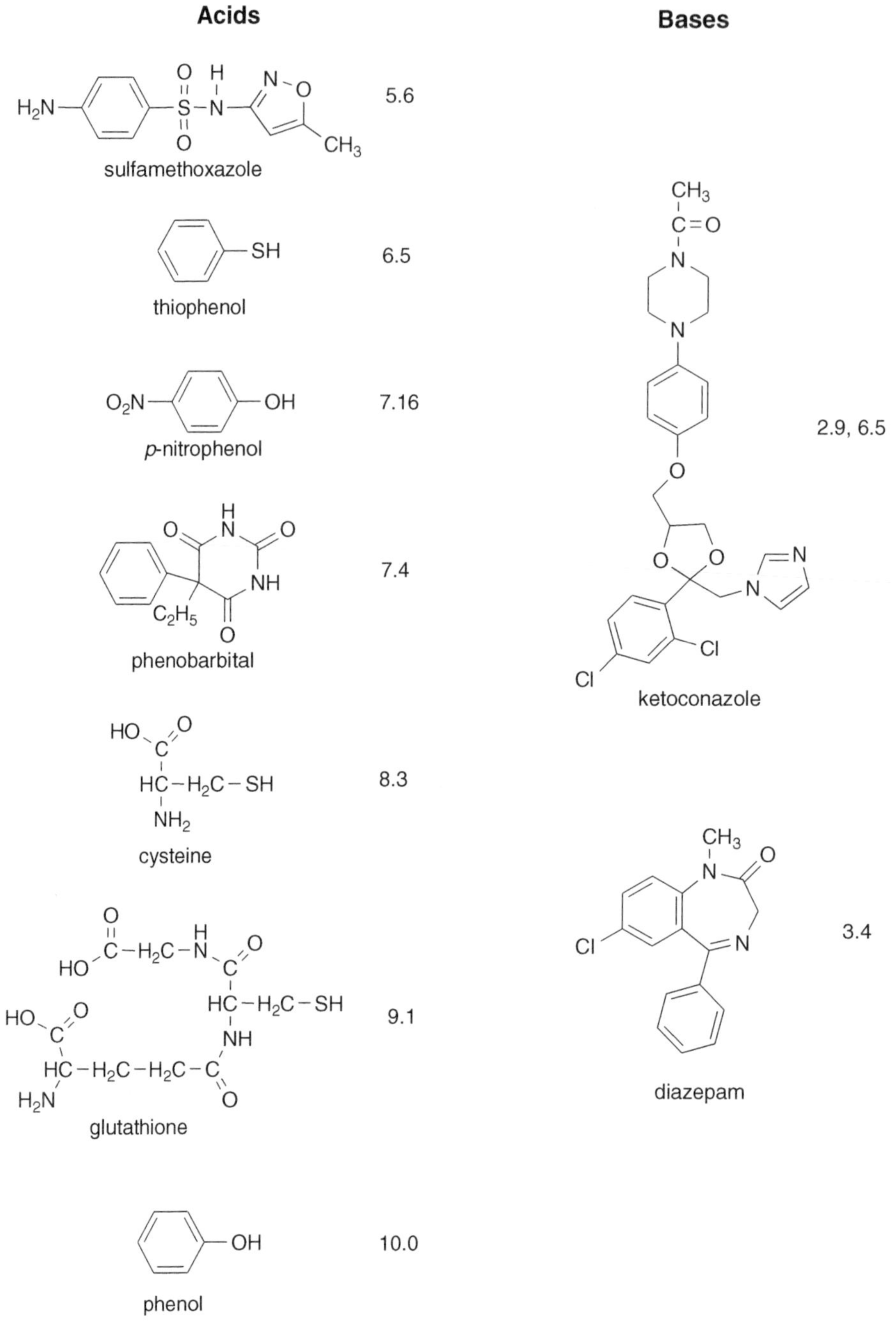

Acids

Bases

sulfamethoxazole 5.6

thiophenol 6.5

p-nitrophenol 7.16

phenobarbital 7.4

cysteine 8.3

glutathione 9.1

phenol 10.0

ketoconazole 2.9, 6.5

diazepam 3.4

FIGURE 2.6 (*continued*)

the arithmetic sum of the two, 6, is called the isoelectric point and is the pH at which the protonated amino group is exactly counterbalanced by the anionic carboxyl group. An isoelectric point with a pH of approximately 6 applies to all the simple amino acids but clearly not to the dibasic (e.g., lysine) or diacidic (e.g., glutamic acid) amino acids (Fig. 2.7). At pHs higher than 6 the concentration of the anionic carboxyl group dominates,

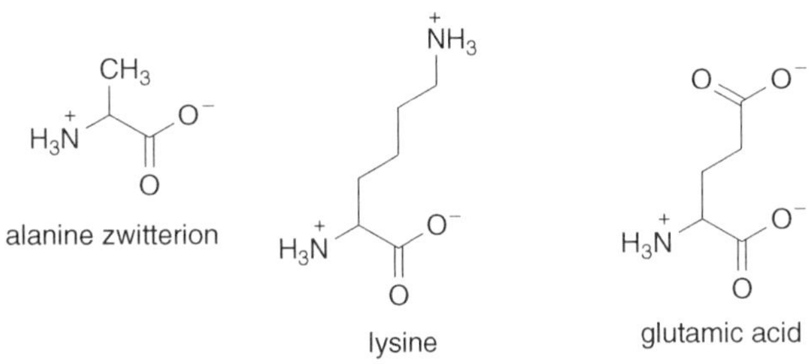

FIGURE 2.7 Structures of "neutral," basic, and acidic amino acids.

while at pHs lower than 6 the concentration of the protonated amino group dominates. Therefore, at physiological pH of 7.4 alanine will exist predominantly in the anionic form. In terms of the polarity of the molecule, the presence of multiple acidic and/or basic groups increases polarity rather than "neutralizing" it, although intramolecular interactions between anions and cations can somewhat decrease their interactions with water.

FACTORS SUCH AS CHARGE THAT AFFECT DRUG ABSORPTION, DISTRIBUTION, AND EXCRETION

As stated above, charge has a dramatic effect on the properties of a drug. Although there are exceptions, most drugs or drug metabolites for which renal clearance is dominant are usually charged. In contrast, the type of molecule that is well absorbed and passes through the blood–brain barrier is usually lipophilic because it has to pass through a lipid cell membrane. However, this is an oversimplification and the structures shown in Figure 2.8 demonstrate the limitations to this concept.

Charged drugs or other molecules do not readily pass through lipid membranes. However, this often leads to the mistaken conclusion that drugs that are strong acids or bases, i.e., mostly ionized at physiological pH, will not be absorbed from the gut or pass the blood–brain barrier. If a drug is a strong base and has a pK_a of 9.4, the ratio of ionized to un-ionized will be 100:1 at pH 7.4, but that still means that approximately 1% of the drug is un-ionized and free to go through a lipid membrane. And there is always about 1% that is un-ionized because the equilibrium is virtually instantaneous; therefore, the effect of the drug being a strong base or acid is that it decreases the effective concentration available to diffuse across the membrane. This slows the rate of diffusion but not the ultimate amount absorbed given adequate time for absorption. With absorption from the gut, the surface area is quite large and transit time in the small intestine, where most absorption occurs, is usually more than 4 hours. In addition, there is a significant gap between cells so that small molecules do not have to go through a lipid membrane. In fact, the rate of absorption from the gut for such molecules is usually based principally on the rate of gastric emptying, and the fact that a drug has one strong acid or base does not usually significantly affect the rate of absorption. In contrast, the surface area of the blood–brain barrier is less than that of the gut and the gap between cells is tight. Cocaine is a good example to see the effect of a basic or acidic group. Using the principles from the previous section, if you look at the structure of cocaine (Fig. 2.8), there is a basic nitrogen that is mostly charged at physiological pH. However, even though it is mostly ionized, cocaine not only gets through the blood–brain barrier into the brain but gets through very rapidly. In fact, almost all drugs of abuse and other drugs active in the central nervous system (CNS) are basic amines mostly charged at physiological pH. This includes cocaine, heroin, amphetamines, phencyclidine, nicotine,

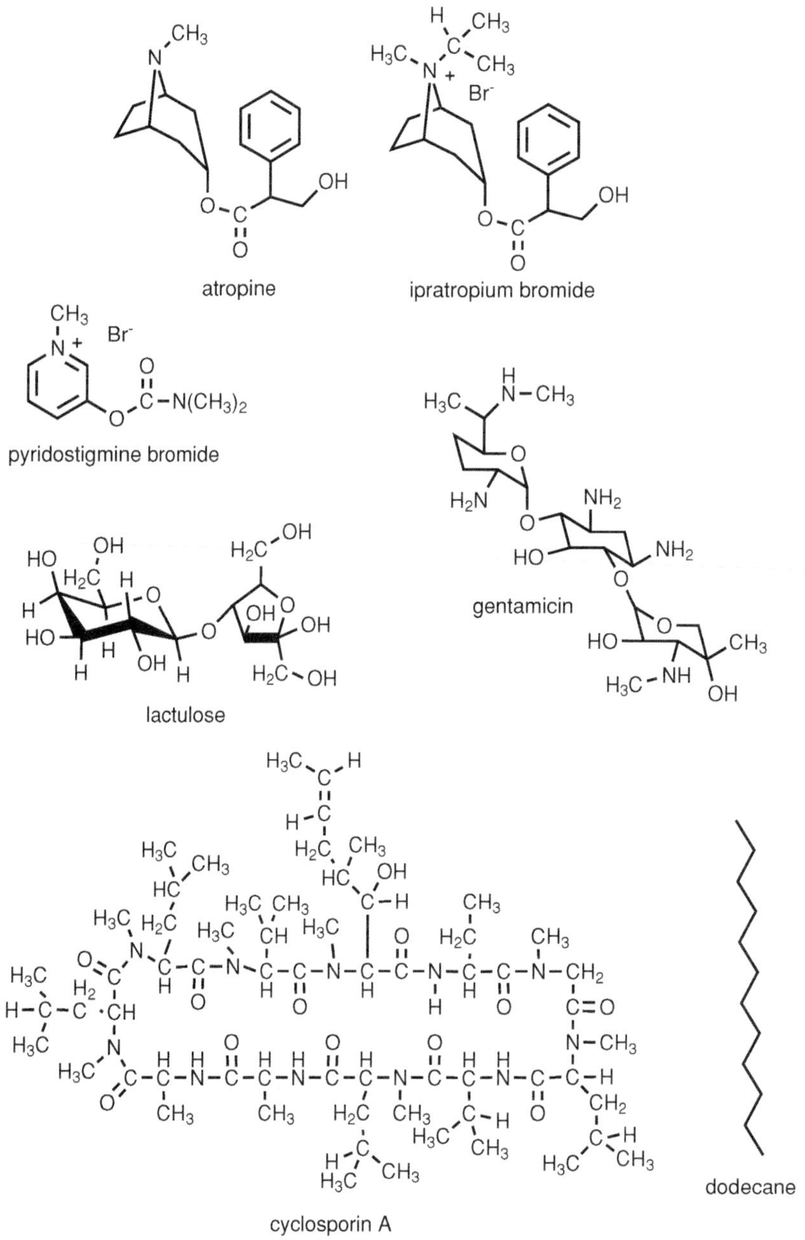

FIGURE 2.8 Examples of molecules to illustrate features that determine absorption and penetration of the blood–brain barrier.

antidepressants, and antipsychotics. It is possible that there are transporters for some of these drugs that facilitate penetration of the blood–brain barrier, but it is unlikely that this is responsible in all or even most cases. There are however limits, and drugs in which there are several acidic and/or basic groups or a very highly ionized functional group such as a sulfonic acid (see azetreonam above) do not penetrate the blood–brain barrier to a significant degree in the absence of a transport system.

In contrast to tertiary amines, drugs that are quaternary ammonium salts are 100% ionized, not well absorbed from the gut, and, in general, do not have CNS effects. For example, atropine, which has a structure similar to cocaine (Fig. 2.8), is an anticholiner-gic drug that has bronchodilitory effects and has been used for the treatment of asthma; however, its usefulness is limited due to CNS side effects. In order to avoid this problem, ipratropium bromide was developed in which the terteriary amine has been converted to a quaternary ammonium salt and therefore is 100% ionized (Fig. 2.8). This drug can be inhaled and can exert its therapeutic effects in the lungs, but it cannot penetrate the blood–brain barrier and thus is devoid of CNS side effects making it a much more useful drug for the treatment of asthma. However, when an anticholinergic agent is used for the treatment of organophosphate poisoning, it is important that it get into the brain because much of the toxicity of organophosphates is due to increased concentrations of acetylcholine in the brain. Therefore, atropine is the agent of choice for the treatment of organophosphate poisoning rather than ipratropium bromide. Even though they are 100% ionized, the oral bioavailability of small quaternary drugs such as pyridostigmine, which is administered orally for the treatment of myasthenia gravis, is low but significant (about 10%). This is probably because they are able to go through the gaps between cells in the gut. This is also true of other drugs such as the bisphosphonates that are highly charged.

As mentioned above, one acidic or basic functional group will not keep a drug from passing through a lipid membrane as long as a significant fraction of the acidic or basic group is un-ionized; however, the presence of many such groups will. For example, gentamicin, which is an aminoglycoside antibiotic, has five basic amino groups, and the probability that all will be un-ionized simultaneously is quite low. Therefore, the oral absorption of gentamicin is low and it must be given parenterally to achieve therapeutic blood concentrations. However, absorption is not zero and there are cases in which oral aminoglycosides given to "sterilize" the gut, especially if they were retained in the gut for long periods of time, have caused partial deafness because of their ototoxicity. In fact, even the presence of many OH groups, which are not charged, is sufficient to prevent absorption. The only reason that most sugars such as glucose are well absorbed is that there is a transport system to facilitate their absorption. In contrast, lactulose (Fig. 2.8), which is not a substrate for such transporters, is not well absorbed and is used as a laxative. This is presumably because each of the OH groups is associated with waters of hydration and it requires a large amount of energy to break these interactions as the molecule enters a lipid membrane. In general, peptides also have poor bioavailability not only because they are hydrolyzed by peptidases in the gut but also because they contain many peptide bonds (amide linkages), which form hydrogen bonds to water.

Lipinsky proposed the rule of 5; specifically, if a drug has more than 5 hydrogen bond donors (N–H or O–H groups), more than 10 hydrogen bond acceptors (sum of nitrogen and oxygen atoms), a molecular mass of more than 500, or a log P (logarithm of the octanol/water partition ratio) greater than 5, it is likely to have poor bioavailability (4). The rule of 5 is only a rough guide; there are exceptions and it is more complex than this (5). One exception is cyclosporin, which is a polypeptide with a high molecular weight. However, most of the amide bonds in cyclosporin are methylated, which prevents hydrogen bonding. Those that are not methylated are involved in intramolecular hydrogen bonds, which reduce their interaction with water. Figure 2.8 shows the N–H bonds that are close to carbonyl groups on the other side of the ring involved in intramolecular hydrogen bonds and also shows the methylated peptide linkages. The result is that cyclosporin has reasonable oral bioavailability, and a major factor that limits its bioavailability is the fact that it is a sub-strate for the transporter, P-glycoprotein, which keeps pumping it back into the gut lumen.

It might be inferred that lipophilic molecules would be well absorbed, but there is also a limit at the other extreme. The 12-carbon chain alkane (dodecane, a representative of mineral oil, which is actually a mixture of alkanes of approximately this size) shown in

Figure 2.8 is not well absorbed and mineral oil has also been used as a laxative. Although such molecules go into lipid membranes, they cannot "get out the other side," i.e., enter the blood stream, because their water solubility is very low. Solids with very low water solubility have even more difficulty because solids cannot interact with lipid membranes as well as liquids.

Although the concepts presented in this section are useful for predicting passage through membranes, a major complicating factor is the presence of transporters. The transporter field is growing rapidly and is beyond the scope of this book.

HINTS FOR UNDERSTANDING CHEMICAL MECHANISMS

In the following chapters, there will be many examples where mechanisms of reactions will be illustrated with arrows showing the movement of electrons (that result in the breaking of old bonds and the making of new bonds) as bonds are formed and broken. It will be easier to understand these mechanisms if it is kept in mind that most chemical reactions can be viewed as the reaction of electrophiles (molecules that are electron deficient) with nucleophiles (molecules that have an atom with a relatively high electron density). Therefore, electrons move from nucleophiles or a negatively charged atom toward electrophiles or a positively charged atom or, if there is no charge, electrons move toward the more electronegative of the two atoms involved. In the process, the correct number of bonds to each atom should be kept appropriate. Therefore, if you draw a mechanism that has electrons flowing away from a positive charge or forming five bonds to a carbon, it is unlikely to be correct. In addition, it is easy to lose or gain a proton in an aqueous solution; in contrast, the gain or loss of a hydride (negatively charged hydrogen) is rare and occurs only with the help of a specific enzyme and hydride acceptor/donor. The gain or loss of a neutral hydrogen atom implies that the reaction involves free radical chemistry.

Hydrogen atom abstraction from a molecule has been proposed as an intermediate during the oxidation of C–H bonds by cytochromes P450. Hydrogen atom abstraction can also occur during oxidative stress when reactive oxygen–free radicals are formed. An arrow representing the movement of electrons with two barbs represents the movement of two electrons while an arrow with only one barb represents the movement of only one electron. If a bond is broken by moving two electrons in one direction, it will result in a positive charge on the tail of the arrow and a negative charge at the head of the arrow. These charges are often dissipated by the loss or gain of a proton, i.e., the loss of a proton will remove the positive charge on a molecule while gaining a proton will eliminate a negative charge on a molecule. However, if a bond is broken by a single electron leaving that bond, no charge is produced but rather free radicals are generated. This is symbolized by an arrow with a single barb indicating the direction of electron movement. The head of the single-barbed arrow indicates a single electron and the tail of the arrow indicates the single electron left behind. Examples that you will come across later in this book are shown in Figure 2.9.

The first example shows the rearrangement of an arene oxide (epoxide). Because of bond-angle strain, the bond between the carbon and oxygen breaks and the electrons go to the more electronegative oxygen atom. This results in a negative charge on the oxygen and a positive charge on the *m* carbon with overall conservation of charge. A hydride shifts over to the carbocation to neutralize the charge. (This hydride is not free in solution; rather it is the hydrogen with its electrons moving over to the adjacent carbon to neutralize the positive charge and the new C–H bond is formed simultaneously with breaking the old C–H bond.) This would leave the *p* carbon atom with only three bonds, but simultaneously the electrons from the oxygen form a double bond to this carbon and it is the availability of

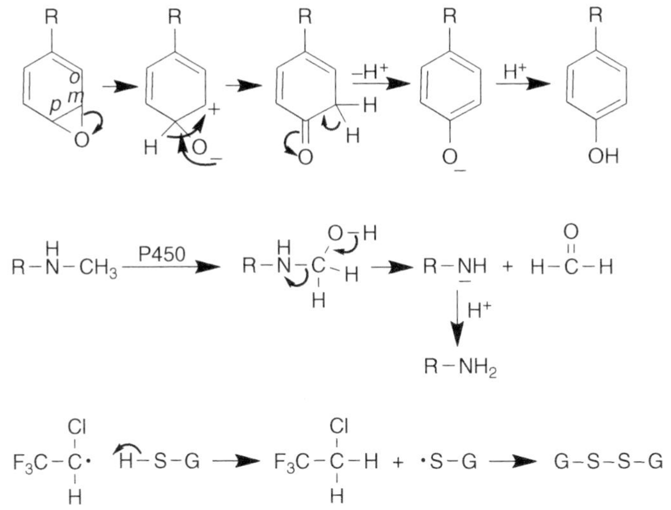

FIGURE 2.9 Examples of reaction mechanisms in which arrows show the movement of electrons.

these electrons that facilitates the hydride shift. In the next step, the electrons from one of the two *m* C–H bonds form a C–C bond in the ring thus regenerating an aromatic system along with the release of a proton. In order to keep the appropriate number of bonds to the *p* carbon, the electrons from one of the C–O bonds move to the electronegative oxygen forming an anion. This anion picks up a proton from the environment to neutralize this charge leading to the final product.

In the second example, N-dealkylation, oxidation of the carbon next to the nitrogen leads to a carbinolamine. This spontaneously leads to formaldehyde and an amine. The mechanism involves loss of a proton with electrons moving toward the electronegative nitrogen atom. The negative charge on the nitrogen is neutralized due to the addition of a proton from the environment.

In the last example, the free radical formed from the reductive dehalogenation of halothane abstracts a hydrogen atom from glutathione resulting in a more stable glutathione-free radical (glutathione is a tripeptide abbreviated as G-S-H to emphasize the sulfhydryl group, which is the site of most glutathione reactions). Although the concentration of free radicals in a biological system is usually quite low, free radicals readily react with other free radicals to form a new bond; in this case, two glutathione-free radicals can react to form the dimer, GSSG. In the case of free radical reactions, as a bond is broken or formed, the two electrons that form the bond are equally distributed to the atoms involved in the bond, and therefore no charge is generated.

The reactions of reactive metabolites can usually be viewed as nucleophilic substitution reactions. Nucleophilic substitution reactions are separated into S_N1 and S_N2 reactions. In an S_N1 reaction, there is usually a negatively charged leaving group, which leaves behind a positively charged molecule, often a positively charged carbon called a carbocation, that can react with a nucleophile. At reasonable concentrations, the rate of the reaction depends only on the concentration of the compound with the leaving group; therefore, the reaction is referred to as a substitution reaction, nucleophilic, first order because the reaction rate depends only on the concentration of one molecule. Specifically, the rate of the reaction depends on the rate of dissociation of the molecule to form a carbocation or other cation. A classical example of an S_N1 reaction is the rate-determining formation of a benzylic carbocation from benzyl bromide, followed by the very rapid reaction of the carbocation

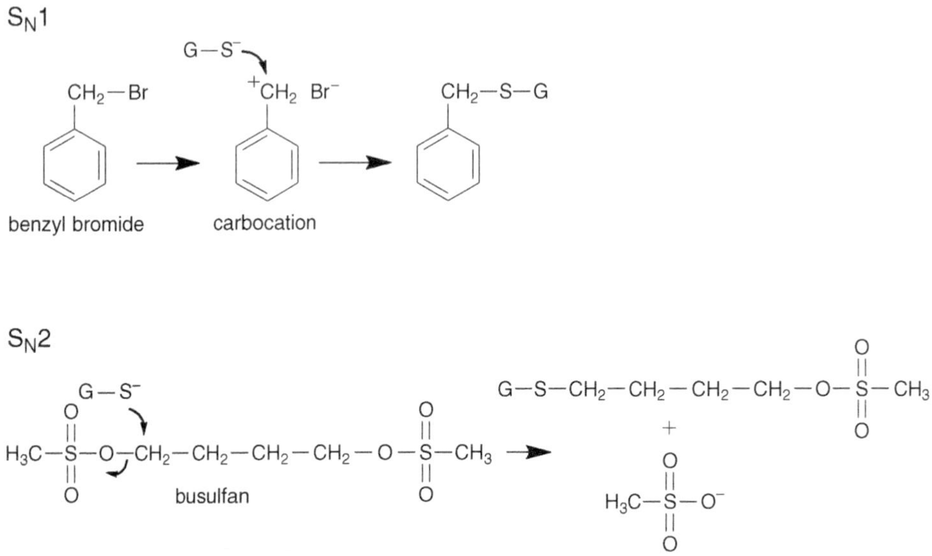

S_N1

S_N2

FIGURE 2.10 Differentiation of S_N1 (substitution nucleophilic unimolecular, first order) and S_N2 (substitution nucleophilic bimolecular, second order) reactions.

with glutathione as shown in Figure 2.10. Even more stable cations are formed from drugs such as clozapine as discussed in Chapter 5.

In contrast, an S_N2 reaction involves a nucleophile attacking a carbon. Formation of the bond to the nucleophile occurs simultaneously with the breaking of the bond to the leaving group. A classic example is the reaction of glutathione with busulfan (Fig. 2.10), which would have to form a primary carbocation if it were an S_N1 reaction. Another example is 2,4-dinitrofluorobenzene, which is very electrophilic, but the carbocation that would have to be formed in an S_N1 reaction would be extremely difficult to form because it involves a sp^2-hybridized carbon and it is also very electron deficient because of the nitro groups. Because the rate of the reaction is dependent on the concentration of both the nucleophile and the molecule being attacked, it is a second-order reaction or an S_N2 reaction. It may be necessary to review this concept with the help of a basic chemistry textbook. In addition, it will be helpful as you go through this book to see how these concepts apply to each of the mechanisms because it is only with practice that you will become comfortable with them and write mechanisms of your own.

REFERENCES

1. Perrin DD, Dempsey B, Serjeant EP. pK_a Pediction for Organic Acids and Bases. London: Chapman and Hall; 1981.
2. Foye WO, Lemke TL, Williams DA. Principles of Medicinal Chemistry (4th ed). Media, PA: Williams & Wilkins; 1995.
3. Budavari Se. The Merck Index (12th ed). Whitehouse Station: Merck Research Laboratories; 1996.
4. Lipinski CA, Lombardo F, Dominy BW, et al. Experimental and computational approaches to estimate solubility and permeability in drug discovery and development settings. Adv Drug Delivery Rev 1997;23:3–25.
5. Stenberg P, Bergstrom CA, Luthman K, et al. Theoretical predictions of drug absorption in drug discovery and development. Clin Pharmacokinet 2002;41(11):877–899.

3
Background for Chemists

BASIC ASPECTS OF PHARMACOKINETICS

Because the chemical structure of a molecule encodes its biological properties, structure has long served as the primary variable and determinant for the discovery of new drugs by medicinal chemists. For this reason, systematic structural modification has been the primary tool of choice to isolate and enhance a desired biologic activity. Moreover, with the relatively recent development of in vitro receptor-binding assays, combinatorial methods of chemical synthesis, and computer graphics, the overall approach to structural modification has become increasingly sophisticated.

Despite the primacy of structure, it is not the only parameter governing biologic activity. It does not matter how inherently potent a molecule is—if it cannot reach the site of action in sufficient concentration to exert the desired effect, it is essentially useless and will never become a viable medicinal agent. Thus, concentration from a practical perspective is as important as structure. But, in the effort to maximize structural parameters to yield molecules that are inherently ever more potent, an appreciation for the importance of concentration of the agent at the biological site of action has often been lost.

At a fundamental level, the importance of concentration and elaborating the parameters that govern the concentration and lifetime of a drug in vivo, from its adsorption, accumulation, and elimination, has been the focus of the study of the field of pharmacokinetics and its parent discipline biopharmaceutics. More explicitly, pharmacokinetics has been defined as the study and characterization of the time course of drug absorption, distribution, metabolism, and excretion, and the relationship of these processes to the intensity and time course of therapeutic and toxicological effects of drugs. The more inclusive discipline of biopharmaceutics has been defined as the study of the factors influencing the bioavailability of a drug in man and animals and the use of this information to optimize pharmacologic or therapeutic activity of drug products in clinical application (1,2). A major distinction between biopharmaceutics and pharmacokinetics is that biopharmaceutics embraces the science of dosage form development and production whereas pharmacokinetics is limited to the in vivo time course of drug concentration that a given dosage form produces once it is developed.

Bioavailability is a term used to indicate both the rate and fraction of an administered drug that reaches the general circulation intact. The preeminent variable governing bioavailability is the mode of administration. Direct injection into the blood stream or IV infusion results in 100% bioavailability because the drug is placed directly into the systemic circulation without any intervening barriers. In contrast, in all other modes of administration, e.g., subcutaneous injection, sublingual, inhalation, or oral, the drug has to pass through membranes before reaching the systemic circulation. As a consequence, bioavailability can be, and often is, less than 100%. This is particularly true of oral administration where the drug is usually given in solid form as a tablet or capsule. In this mode of administration, before the drug can even be absorbed by gut membranes, the physical form of the drug, tablet or capsule, must first disintegrate and dissolve in the acidic environment of the stomach or the more basic environment of the small intestine. From there, the drug is transported by the portal vein to the liver before finally entering the general circulation. At each stage of this process, a fraction of the drug can be lost. If disintegration is incomplete or the drug does not dissolve, it will not be absorbed and simply pass through the system. Once entering the gut membrane it will encounter P-glycoprotein (P-gp), an energy-dependent transporter system that can act as an efflux pump that, depending on the drugs structure, can actively transport the drug out of the membrane and back into the gut. If it is not expelled back into the gut and continues passage through the membrane, it will encounter the oxidative cytochrome P450 enzymes, particularly CYP3A4, before entering the portal vein. Thus, even before reaching the liver, an orally administered drug is potentially subject to elimination and metabolic transformation. If the drug escapes the upper GI tract and descends further into the gut, it will encounter the gut flora. Gut bacteria, being anaerobic, have a capacity for reducing foreign molecules. For example, nitro-containing drugs, sulfoxides, and the cardiac glycoside, digoxin, are all subject to reductive metabolic transformation. Once in the liver, the drug is subject to the full complement of metabolic enzymes before it enters the general circulation. The fraction of an orally administered drug that is lost because of incomplete absorption and/or metabolic transformation before it enters the general circulation is termed the first-pass effect. So the final fraction of orally administered drug that actually enters the systemic circulation, i.e. the bioavailability, reflects the sum of loss due to incomplete absorption plus the loss due to the first pass effect. Because of these two factors it is not unusual for a drug to be less than 50% bioavailable.

As a drug is absorbed and enters the general circulation, its concentration in the blood begins to rise. Experimentally, for ease of analysis a drug's concentration in blood is measured as its concentration in plasma, i.e., a blood sample is taken, centrifuged to precipitate the red blood cells, and then the supernatant or plasma is analyzed for drug concentration. Plasma concentration is usually not identical to blood concentration as drug will distribute to the red blood cells, and in addition to the fraction dispersed in intracellular fluid, an additional amount may be bound to red blood cell protein or other cellular elements. The fraction associated with red blood cells is not often measured because the drug concentration present in vascular fluid, i.e., plasma, is reflective of the total amount of drug in the body and is an indirect measure of the amount of drug available to the bioreceptors responsible for the biologic effects of that drug.

Upon absorption, the plasma concentration of the drug continues to rise until it reaches the maximum concentration. C_{max}. At C_{max}, the rates of elimination processes such as metabolism and excretion, which also begin to operate on the drug as soon as it enters the body, equal the rate at which it is absorbed (Fig. 3.1). Throughout the absorption process, the drug rapidly distributes to the red blood cells, organs, and all intra- and extracellular

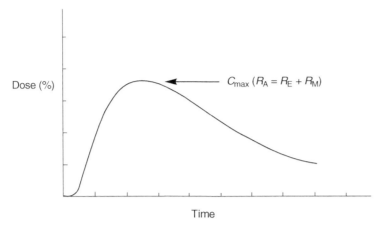

FIGURE 3.1   Absorption curve for an orally administered drug. $C_{\max}$, maximum concentration; R_A, rate of absorption; R_E, rate of elimination; and R_m, rate of metabolism.

sites throughout the body. It drives to establish equilibrium between the drug, body water, proteins contained in body fluids, as well as other cellular components such as lipoidal membranes and macromolecular carbohydrates. Distributive processes are generally much faster than those of absorption or elimination. When the overall rate of elimination exceeds the rate of absorption, the plasma concentration of the drug begins to fall and continues until all the drug is eliminated (Fig. 3.1). The plasma-drug concentration/time curve is the experimental parameter of primary importance to the pharmacokineticist. It contains all the information necessary for the determination of a drug's therapeutic effectiveness, its likely toxicity, and how often to administer the drug. It is therefore important for the medicinal chemist to be familiar with concepts such as drug half-life ($t_{1/2}$), volume of distribution (V_D), clearance (CL), etc., concepts that are the stock in trade of pharmacokineticists because such factors can be decisive in determining a drug's ultimate utility.

After $C_{\max}$ and distribution equilibrium have been reached, the subsequent drug elimination phase can generally be described by first-order kinetics. The time-dependent decrease in drug-plasma concentration is paralleled by a corresponding decrease in elimination rate. Under these conditions, the plasma concentration of the drug at time t is given by Eq. (3.1).

$$\log C = \log C_0^* - kt/2.303 \tag{3.1}$$

In Eq. (3.1), C is the concentration of drug in the plasma at time t, C_0^* is the concentration of drug in the plasma extrapolated to $t = 0$, and k is the first-order rate constant. A semilogarithmic plot of the log C versus t yields a straight line, where the slope of the line is given by $-k/2.303$ and C_0^* is given by the intercept of the y-axis (Fig. 3.2). The first-order rate constant, k, can be determined from the slope of the line or more simply from the relationship stated in Eq. (3.2).

$$k = 0.693/t_{1/2} \tag{3.2}$$

The half-life of a drug, $t_{1/2}$, is simply the time it takes the plasma concentration of that drug to fall to half of its maximum concentration. It is an important parameter because it is a direct measure of body's exposure time to the drug and therefore of the persistence of the biological response caused by the drug. The relationship between $t_{1/2}$ and biologic response makes $t_{1/2}$ particularly useful in determining the frequency of dosing likely to achieve and maintain the desired therapeutic response. While most drugs obey first-order

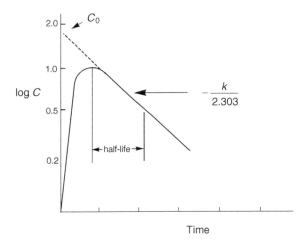

FIGURE 3.2 Semilog plot of the plasma concentration, C, of a drug as a function of time, where C_0 is the projected plasma concentration at time zero and $-k/2.303$ is the slope of the elimination curve.

kinetics because therapeutic doses are generally low enough to allow this to be true, a few do not. Those that do not obey first-order kinetics, e.g., phenytoin and salicylate, can be particularly difficult in finding and maintaining a dose level that is therapeutic without being toxic.

Volume of distribution, V_D, is the proportionality constant that equates the amount of drug in the body (A) to its concentration in plasma (C) [Eq. (3.3)].

$$V_D = A/C \tag{3.3}$$

From Eq. (3.3), V_D is the apparent volume that a given amount of drug will occupy based on its plasma concentration. In and of itself V_D has no physiologic meaning. Since values of V_D can vary from a few liters to a few hundred liters in a 70-kg man, it is clearly not a measure of plasma volume (0.04 L/kg), and hence the term apparent volume. It is primarily a characteristic of the specific drug and reflects drug distribution and binding in the biologic system as a whole. High values result when the drug preferentially distributes to extravascular sites or compartments, while low values reflect preferential confinement of distribution to the blood or vascular compartment. V_D is a useful parameter precisely because it relates the total amount of drug in the body to the plasma concentration and indicates the relative amounts of drug in the vascular and extravascular compartments.

Use of Eq. (3.3) to calculate V_D does not provide an accurate result. The calculated value is invariably high as it assumes immediate distribution of the drug, no metabolic turnover, and no excretion. The assumption is that A is a constant and does not change in the time it takes to measure C. But A is subject to metabolism and excretion from the moment it enters the system, which means that C will be smaller than it would have been in the absence of metabolic turnover. Therefore, if A is divided by a smaller number than it should have been V_D will be too large. Even if the drug is administered intravenously, distribution still takes a finite period of time and metabolism and excretion do occur. Good measures of V_D can however be obtained from Eq. (3.4) for rapidly distributing drugs.

$$V_D = \text{dose}_{IV}/(\text{AUC})k \tag{3.4}$$

In Eq. (3.4), dose_{IV} is the amount of drug administered intravenously, AUC is total area under the drug concentration–time curve, and k is the first-order elimination rate constant

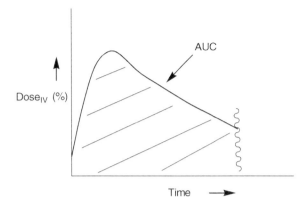

FIGURE 3.3 Area under the plasma decay curve, AUC, after an IV dose of drug.

(Fig. 3.3). V_D can often be determined from IM injection of the drug in lieu of IV administration in which case $dose_{IV}$ is replaced by $dose_{IM}$ in Eq. (3.4). V_D is usually not determined from oral administration and Eq. (3.4). However, if it is, the dose parameter in Eq. (3.4) must be replaced by $dose_{oral}$. $Dose_{oral}$ is a measure of the amount of drug that reaches the systemic circulation and is not simply the amount of drug physically ingested.

The terminator of drug action is, of course, elimination. Elimination is a composite of excretion (kidney, etc.) and biotransformation (metabolism). The primary measure of drug elimination from the whole body is clearance, CL_T, defined as the volume of plasma fluid removed of drug per unit time. It is a direct measure of the loss of the drug from the system and can be calculated from Eq. (3.5) after IV administration of a dose of the drug.

$$CL_T = dose_{IV}/AUC \tag{3.5}$$

CL_T can also be calculated from an oral dose of the drug provided the fraction, f, that is absorbed and reaches the systemic circulation unchanged is also known [Eq. (3.6)].

$$CL_T = f \times dose/AUC \tag{3.6}$$

The clearance of drug from the plasma is a summation of all the clearances of the drug that result on circulation of the drug through the body and through various organs, particularly the liver and kidney. Often it is useful to focus on the clearance of a specific organ. For example hepatic clearance, CL_H, can be determined from the difference in the amount of drug entering the liver to the amount exiting the liver. The amount of drug going into the liver per unit time would be equal to blood flow, Q, times the drug's concentration in arterial blood, C_A. The amount of drug coming out of the liver per unit time would be equal to Q times the drug's concentration in venous blood, C_V. The difference between the two is the portion of the overall rate of elimination of the drug from the blood due solely to the liver [Eq. (3.7)].

$$\text{Elimination rate} = QC_A - QC_V = Q(C_A - C_V) \tag{3.7}$$

The ratio of the elimination rate to the rate of drug input, QC_A, is called the extraction ratio (ER) and is simply the fraction of drug emerging from the liver [Eq. (3.8)]. Hepatic clearance, CL_H, is defined as the volume of blood passing through the liver that is cleared of drug per unit time and is the proportionality constant that relates the elimination rate of the drug from the liver to the concentration of drug in arterial blood that enters the liver [Eq. (3.9)]. As seen from Eq. (3.9), CL_H is also equal to the product of blood flow and the extraction ratio. CL_H is an important parameter because it is a direct measure of the overall enzyme activity in the liver responsible for that drug's metabolism.

$$ER = Q(C_A - C_V)/QC_A = (C_A - C_V)/C_A \qquad (3.8)$$

$$CL_H = Q(C_A - C_V)/C_A = Q(ER) \qquad (3.9)$$

In general, CL_H is the major parameter governing drug elimination and, therefore, one of the primary parameters governing both the magnitude and extent of drug activity. This does not mean that all enzyme activity involved in drug metabolism is localized in the liver, but it does mean that the bulk of drug-metabolizing activity for most drugs is localized in the liver.

A number of different enzyme systems contribute to drug metabolism. Together they embrace several classes of chemical reactions that include oxidation, reduction, hydrolysis, and conjugation reactions. As mentioned in chapter 1, the first three are further classified as phase I reactions while the last, conjugation reactions, are classified as phase II reactions. Phase I reactions are viewed as reactions that introduce some functionality, e.g., alcohol or acid, into the drug molecule. Upon introduction of the new functional group, the modified drug molecule is often subject to the possibility of further metabolic transformation by a phase II reaction. Phase II, or conjugation reactions, generally involves the coupling of some highly water-soluble species, e.g., glucuronic acid or sulfuric acid, with an appropriate functional group present in the drug molecule. Viewing drug elimination as a defense mechanism, the role of phase I reactions is to provide a means of chemically modifying a foreign substance so that it becomes susceptible to conjugation with some highly polar species via a phase II reaction. Such a reaction sequence yields a metabolic product that is much more water soluble than the parent drug and thus more easily and rapidly eliminated from the body. In their totality, these four classes of enzymatically mediated metabolic reactions can operate on essentially any foreign organic molecule that gains entry to the body.

DRUG TRANSPORTERS

Within the last decade, the active transport of drugs by drug transporters has emerged as an important factor that can modulate drug distribution. While numerous transporters have been identified in a variety of tissues that transport nutritional elements such as carbohydrates, peptides, and minerals, one transporter, P-gp, stands out as particularly important in the transport of drugs (3). P-gp is an ATP-dependent glycoprotein and a member of the large ATP-binding cassette protein transporter family, which has been extensively characterized as an important element of the phenomenon of multidrug resistance that plagues effective cancer chemotherapy.

The scope of this volume does not allow discussion of all the important transporters, including multidrug resistance proteins, organic anion transporting proteins, and organic cation transporting proteins, but they need to be considered. Therefore, P-gp, arguably the most general and important of the drug transporters, will serve as a model for the class of drug transporters in the body with the full realization that it is not the only transporter.

P-gp is expressed in tumor tissue and serves as a barrier to increased accumulation of cytotoxic anticancer drugs within the cancer cell by actively pumping the drug out of the cell. It is a multidrug resistance element by virtue of the fact that it is promiscuous in terms of substrate selectivity. It, like the cytochromes P450, will accept a wide diversity of structural types. For example, a small sampling of drugs that are substrates for P-gp include the anticancer agents, actomycin D, doxorubcin, etoposide, mitomycin C, taxol, vincristine, and other drugs such as cortisol, digoxin, indinavir, morphine, progesterone, and terfenadine.

In normal tissue, P-gp is located in intestine, liver, and kidney and also serves as a significant component of the blood–brain barrier. Its location in intestine has a major influence on drug absorption and bioavailability, particularly in conjunction with CYP3A4, which is also present in intestinal epithelium. As drug is absorbed from the stomach and passes into the intestinal epithelium, it is exposed to both P-pg and CYP3A4. CYP3A4 can metabolize the drug while P-gp serves as a barrier to limit both absorption and bioavailability by actively pumping drug out of intestinal epithelial cells back into the lumen. If drug is reabsorbed, it must again face both metabolism by CYP3A4 as well as rejection by P-gp. In the liver P-gp facilitates the excretion of both the drug and drug metabolites into bile, whereas in the brain it serves as a major barrier to access.

Since P-pg can have such a major influence on bioavailability and distribution and since it has life-threatening implications, particularly with regard to anticancer chemotherapy, the availability of inhibitors that could modulate its effects would be highly desirable. A number of drugs that will, in fact, reverse the effects of P-pg have been found. They include the calcium channel blockers, verapamil and nifedepine, the antiarrhythmic agents, quinidine and amiodarone, as well as a number of other agents. Not surprisingly, the search for potent inhibitors with less pharmacological liability is a very active area of research and potentially effective agents are beginning to emerge.

ENZYME KINETICS

The kinetic behavior of drugs in the body can generally be accounted for by first-order kinetics that are saturable, i.e., Michaelis–Menten kinetics. A brief review of the principles of Michaelis–Menten kinetics is given next (4).

At low substrate concentrations, the rate of reaction is first order, i.e., it is proportional to substrate concentration (Fig. 3.4). As the substrate concentration is increased, the rate begins to fall, i.e., it no longer increases proportionately with increasing substrate concentration. With further increasing substrate concentration the enzyme becomes saturated, the rate becomes essentially constant, and no longer responds to increasing substrate concentration.

At the simplest level, the process can be considered as taking place in two steps. In the first step, enzyme (E) combines reversibly with substrate (S) to form an enzyme–substrate complex (ES) [Eq. (3.10)]. In the second step, ES breaks down to form free E and product (P) [Eq. (3.11)]. This process is also considered to be reversible as indicated by the various rate constants (k) for both forward and reverse reactions.

$$\text{E} + \text{S} \underset{k_2}{\overset{k_1}{\rightleftharpoons}} \text{ES} \qquad\qquad (3.10)$$

$$\text{ES} \underset{k_4}{\overset{k_3}{\rightleftharpoons}} \text{E} + \text{P} \qquad\qquad (3.11)$$

In the Briggs–Haldane derivation of the Michaelis–Menten equation, the concentration of ES is assumed to be at steady state, i.e., its rate of production [Eq. (3.12)] is exactly counterbalanced by its rate of dissociation [Eq. (3.13)]. Since the rate of formation of ES from E + P is vanishingly small, it is neglected. Equating the two equations and rearranging yields Eq. (3.14), where K_M replaces $(k_2 + k_3)/k_1$ and is known as the Michaelis–Menten

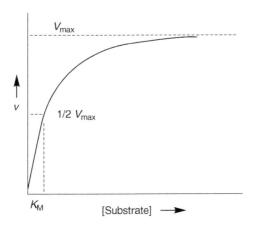

FIGURE 3.4 Plot of the rate of substrate turnover by an enzyme with increasing substrate concentration.

constant. Solving Eq. (3.14) for [ES], the steady-state concentration of the enzyme substrate complex, gives Eq. (3.15). At low substrate concentration the initial velocity (v) is proportional to [ES] as expressed in Eq. (3.16), while the maximum velocity (V_{max}), the velocity when all the enzyme is occupied by saturating concentrations of substrate, is expressed in Eq. (3.17).

$$\frac{d[ES]}{dt} = k_1([E] - [ES])[S] \tag{3.12}$$

$$-\frac{d[ES]}{dt} = k_2[ES] + k_3[ES] \tag{3.13}$$

$$\frac{([E] - [ES])[S]}{[ES]} = \frac{k_2 + k_3}{k_1} = K_M \tag{3.14}$$

$$[ES] = \frac{[E][S]}{K_M + [S]} \tag{3.15}$$

$$v = k_3[ES] \tag{3.16}$$

$$V_{max} = k_3[E] \tag{3.17}$$

Substituting the term for [ES] in Eq. (3.16) into Eq. (3.15), dividing this equation by Eq. (3.17), and then solving for v, the Michaelis–Menten equation, Eq. (3.18) is obtained.

$$v = \frac{V_{max}[S]}{K_M + [S]} \tag{3.18}$$

This equation defines the quantitative relationship between the substrate concentration and enzyme reaction rate when the constants, V_{max} and K_M, are known. An interesting and important relationship emerges when v is equal to $1/2V_{max}$. Under these conditions, [S] is equal to K_M.

Since the therapeutic dose of most drugs generally puts them in the realm of first-order kinetics, Michaelis–Menten kinetics apply. Michaelis–Menten kinetics is especially valuable in drug metabolism studies as an experimental technique to define the various metabolic pathways, particularly the P450-catalyzed metabolic pathways, which are likely to control the metabolism of the drug under consideration. With all the major human P450s being commercially available means that for any drug under study, it is relatively straightforward to determine the metabolites formed and their corresponding V_{max} and K_M values for any specific P450. Such a study allows one not only to determine the major P450s contributing to the metabolism of that drug by comparing relative turnover numbers from V_{max}/K_M values, but it also establishes the degree to which various metabolites are produced by more than one P450. Thus the dominant enzymes controlling the metabolism of the drug under investigation, and in a larger sense the lifetime of its therapeutic effectiveness, are determinable. This approach assumes, of course, that the in vitro results obtained from cloned human enzymes reflect, at least in a general sense, what will happen in vivo. Results suggest that the assumption is largely true and well justified.

A second major use of cloned human P450 enzymes is to determine the ability of a drug to inhibit a P450. This of course has major ramifications on the potential for drug interactions. In multidrug therapy, which is generally the norm in an aging population, if two drugs interact with the same enzyme, the kinetics of either one or both the drugs might be affected and thereby alter the expected metabolic profile. An altered metabolic profile in this context means that the patient would be exposed to enhanced levels of one or both the drugs for an extended period of time, and this in turn might result in adverse consequences. Thus, determination of the enzymes that a drug will interact with, either as a substrate or an inhibitor, is an important goal for good therapeutic management.

If a drug is a substrate of an enzyme, it will also be a competitive inhibitor of that enzyme, but it may be a competitive inhibitor without being a substrate. This is because the rate of product formation is determined by k_3 of the Michaelis–Menten equation while the rate of ES substrate dissociation and degree of enzyme inhibition is determined by the ratio of k_2/k_1 as discussed above. If k_3 is very small it will not be experimentally measurable; however, the enzyme will still be bound and occupied as determined by k_2/k_1.

The ability of being able to determine the enzyme inhibitory profile of a potentially new drug substance by in vitro experimentation has provided the pharmaceutical industry with a powerful new tool for assessing the potential drug interaction liabilities of the new drug before it is ever administered to a living, breathing human being. That means, at least in some cases, serious problems that might arise much later in the new drug's development phase could be avoided before it ever reached this stage. Early recognition and termination of the development of a potentially problematic drug would provide major benefits in human well-being as well as financial cost.

As discussed above, the degree of inhibition is indicated by the ratio of k_2/k_1 and defines an inhibitor constant (K_I) [Eq. (3.19)], whose value reports the dissociation of the enzyme–inhibitor complex (EI) [Eq. (3.20)]. Deriving the equation for competitive inhibition under steady-state conditions leads to Eq. (3.21). Reciprocal plots of $1/v$ versus $1/S$ (Lineweaver–Burk plots) as a function of various inhibitor concentrations readily reveal competitive inhibition and define their characteristic properties (Fig. 3.5). Notice that V_{max} does not change. Irrespective of how much competitive inhibitor is present, its effect can be overcome by adding a sufficient amount of substrate, i.e., substrate can be added until V_{max} is reached. Also notice that K_M does change with inhibitor concentration; therefore the K_M that is measured in the presence of inhibitor is an apparent K_M. The true K_M can only be obtained in the absence of inhibitor.

$$K_I = \frac{k_2}{k_1} \tag{3.19}$$

$$K_I = \frac{[E][I]}{[EI]} \tag{3.20}$$

$$v = \frac{V_{max}[S]}{K_M(1 + [I]/K_I) + [S]} \tag{3.21}$$

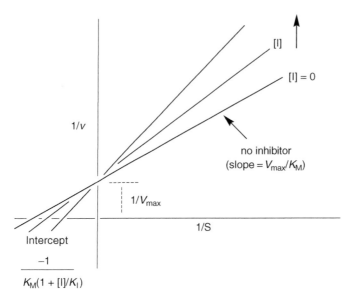

FIGURE 3.5 Reciprocal plot of the rate of substrate turnover by an enzyme in the presence of different concentrations of competitive inhibitor.

A noncompetitive inhibitor is one that inhibits the enzyme and its inhibitory activity is unaffected by substrate, i.e., it will inhibit the enzyme to the same degree whether the substrate is present or not. This is generally thought to occur by the inhibitor binding at some site other than the substrate-binding site but in a way that inactivates the enzyme, e.g., induced conformational change of the active site. Therefore, we may have inhibitor binding reversibly to free enzyme [Eq. (3.22)] or to the enzyme substrate complex [Eq. (3.23)], but in both cases the bound enzyme is inactive.

$$I + E = EI \tag{3.22}$$

$$I + ES = ESI \tag{3.23}$$

A reciprocal plot of the effect of varying concentrations of a noncompetitive inhibitor on enzyme-catalyzed substrate turnover readily reveals the nature and characteristics of this type of inhibition (Fig. 3.6). Notice that in this case, the properties that characterize noncompetitive inhibition are virtually opposite of those that characterize competitive inhibition. With a noncompetitive inhibitor V_{max} does change but K_M is constant.

A special case of noncompetitive inhibition that is sometimes seen in drug metabolism studies, particularly with cytochrome P450-catalyzed reactions, is suicide inhibition. A suicide substrate is one in which during the course of metabolism some fraction of substrate is transformed into a reactive electrophilic intermediate that covalently binds to the active site of the enzyme or the heme cofactor. As a consequence, the enzyme is irreversibly inactivated. Since it is being generated by metabolism, the effect is often time dependent and cumulative. This is in contrast to the effects of competitive inhibition that has a very rapid onset but, in the absence of steady-state conditions, dissipates as the inhibitor itself is metabolized and removed from the body. Clearly, from a therapeutic standpoint, suicide inhibitors or slowly reversible noncompetitive inhibitors are likely to lead to more serious drug interaction problems. Thus, being able to recognize such compounds by in vitro studies early in drug development represents a major advance.

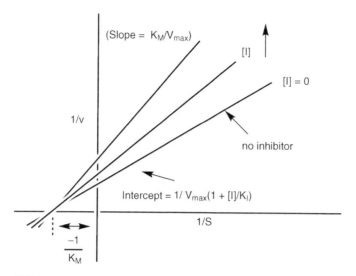

FIGURE 3.6 Reciprocal plot of the rate of substrate turnover by an enzyme in the presence of different concentrations of a noncompetitive inhibitor.

IN VITRO–IN VIVO CORRELATION OF DRUG BEHAVIOR

As discussed earlier, the tacit assumption of in vitro studies is that they are faithful reporters of how the enzymes and substrates will behave in vivo. At least qualitatively, the assumption seems largely to be true but quantitatively the assumption is less reliable. It assumes that the different microenvironments surrounding an enzyme in vivo and in an in vitro preparation do not differentially affect kinetic properties. It also assumes that, given equal concentrations of drug, the concentration that actually reaches the active site of the enzyme in the two different microenvironments will be equal (5). Clearly this does not need to be the case. As a consequence, a more reliable reporter of the in vivo kinetic properties of a drug would be highly desirable.

One of the important kinetic properties that needs to be determined in assessing the drug interaction potential of any drug is the value of its K_I with each of the major cytochrome P450s. The standard equation for determining competitive inhibition of an enzyme is Eq. (3.24). Since a drug does not need to be a substrate for any specific P450 to be a potent inhibitor of that enzyme, its potential for inhibition can only be assessed and evaluated by in vitro studies. Such studies not only help in determining which P450s are susceptible to inhibition but also to what degree. If the inhibitory drug is coadministered with another drug whose metabolism is primarily dependent upon a P450 that is subject to inhibition by the inhibitor, the potential for a serious interaction would be predicted.

$$i = \frac{[I]}{[I] + K_I(1 + [S]/K_M)} \tag{3.24}$$

where i is the fraction of inhibition.

In vivo the corresponding parameter to a K_I is $K_I$iv, a parameter that is determined directly from in vivo experiments. Theoretically, K_I and $K_I$iv should be equal, but for the reasons outlined above they may not be. K_I is a direct measure of the molecular interaction of the drug with the enzyme. $K_I$iv on the other hand is a measure of the actual in vivo effectiveness of the inhibitor. That is $K_I$iv, unlike K_I, automatically incorporates into its value the effects of factors such as differences in active site inhibitor concentration, environmental differences, and/or inhibitor metabolism. Thus $K_I$iv, determined from Eq. (3.25), should be a powerful and practical parameter for assessing the effective inhibitory capacity of drug in an in vivo environment and for confirming an interaction due to the inhibition of a specific P450 identified from previous in vitro studies. Moreover, once the $K_I$iv is determined it should provide a quantitative measure of the degree of inhibition to be expected for any substrate of that enzyme, i.e., $K_I$iv is substrate independent. In Eq. (3.25), [I] is inhibitor concentration at steady state, $CL_{f(c)}$ is the formation clearance (for each metabolite it is calculated according to $CL_{f(c)} = f_m \times CL_T$, where CL_T is the total clearance of the substrate from the body and f_m is the fraction of dose recovered in the urine as a specific metabolite) to the metabolite in the absence of the inhibitor, and $CL_{f(i)}$ is the formation clearance to the metabolite in the presence of inhibitor.

$$K_I iv = [I]/CL_{f(c)}/CL_{f(i)} - 1 \tag{3.25}$$

Fortunately, in order for a metabolically-based drug interaction to become therapeutically significant three criteria must be met (6). First, the substrate drug should have a narrow therapeutic index so that a three- or fourfold increase in its plasma concentration from the norm has significant biologic consequences. If the drug has a wide therapeutic index, a three- or fourfold increase in its plasma concentration would not be toxicologically significant. Second, a single enzyme should account for at least 0.7 of the fraction metabolized, f_m, in the clearance of the drug. If the primary metabolism of a drug is fractioned between a number of different enzymes, inhibition of any one enzyme should not have a major effect because even total inhibition of that enzyme will only account for a relatively small fraction of total clearance of drug. Thus the potential for a dramatic increase in the plasma concentration of the inhibited drug will be severely restricted. Third, the plasma concentration of the inhibitor should be well in excess of its K_I. This insures that the inhibitor can effectively compete with the substrate drug for the enzyme.

While the necessary conditions are restrictive, they can be met as attested to by the drug interactions that do occur. A classic example in this regard is the anticoagulant warfarin. Warfarin is probably more susceptible to clinically significant drug interactions as a result of the co-administration of other medications than any other commonly administered drug. The anticoagulant has a narrow therapeutic index. Too much of the drug can lead to internal bleeding incidents, while too little defeats the intent of therapy to increase clotting time leading to a reduction in the incidence of clot formation. Warfarin is administered as a racemate but (S)-warfarin is the most pharmacologically active enantiomer. A single cytochrome P450, CYP2C9, is responsible for better than 80% of the clearance of (S)-warfarin as the (S)-6- and (S)-7-hydroxywarfarin metabolites. Because all that is required for a significant warfarin drug interaction to occur is inhibition of CYP2C9, it is not surprising that it is highly susceptible to interactions.

DEUTERIUM ISOTOPE EFFECTS

A methodology that has turned out to be a very powerful tool in trying to unravel the intricacies of the mechanism of cytochrome P450-catalyzed oxidation reactions has been the use of deuterium isotope effects. The use of intramolecular deuterium isotope effects have been particularly important in this regard as will be described in chapter 4 where a number of such studies are presented. But, before describing the specific technique that intramolecular isotope effect studies entail, a quick mini review on the nature of deuterium isotope effects is probably in order.

Breaking a chemical bond between atoms involves adding enough energy to the system to increase the vibrational stretching frequency between the two atoms to the point at which the two atoms separate. Because of the twofold difference in mass between deuterium and hydrogen, it is generally more difficult to break a carbon–deuterium bond than it is to break the corresponding carbon–hydrogen bond. This is because the greater the mass of the atoms forming a bond, the greater will be the suppression of the vibrational stretching frequency between the atoms. This means that the resting-state or ground-state energy (zero point energy) of a carbon–deuterium bond is lower than the ground-state energy for a corresponding carbon–hydrogen bond. If both the bonds are vibrationally excited to the breaking point (transition state), the energy (activation energy) required to reach that point will be greater for a carbon–deuterium bond than carbon–hydrogen bond because it starts from a lower point. This translates to chemical reactions and means that a reaction that involves breaking a carbon–deuterium bond will be slower than the same reaction that involves breaking a carbon–hydrogen bond.

A primary isotope effect results when the breaking of a carbon–hydrogen versus a carbon–deuterium bond is the rate-limiting step in the reaction. It is expressed simply as the ratio of rate constants, k_H/k_D. The full expression of k_H/k_D measures the intrinsic primary deuterium isotope for the reaction under consideration, and its magnitude is a measure of the symmetry of the transition state, e.g., $-C\cdots H\cdots O-Fe^{+3}$; the more symmetrical the transition state, the larger the primary isotope effect. The theoretical maximum for a primary deuterium isotope effect at $37°C$ is 9. The less symmetrical the transition state, the more product-like or the more substrate-like the smaller the intrinsic isotope effect will be.

A secondary isotope effect results from modulation of the vibrational frequency of the bond-breaking step as a result of a deuterium atom that is not directly involved in the bond-breaking step, but is adjacent to the bond being broken. As would be expected, its magnitude is much smaller. It rarely exceeds 1.5 and is generally around 1.1 to 1.2. The low value for a secondary deuterium isotope effect does not mean it does not have an impact because the effects are multiplicative. For example, suppose the methyl group of an alkyl side chain of some drug were replaced with a trideuteromethyl group. Oxidation of the trideuteromethyl group to a hydroxymethyl group involves one primary deuterium isotope effect and two secondary deuterium isotope effects. The magnitude of the intrinsic deuterium isotope effect for the oxidation would be equal to PS^2, and if the maximum primary and secondary isotope effects were operative, a deuterium isotope effect as large as 20.25 could be observed.

Rarely is an isotope effect of this magnitude observed, particularly for enzymatically mediated reactions. In fact, observed deuterium isotope effects are generally much smaller, often as small as 2 or 3. This is because of the complexity of enzyme-catalyzed reactions. Such reactions generally involve a number of steps, in addition to the bond-breaking step, which can be at least partially rate limiting, e.g., product release, binding effects, etc. Moreover, the rates of transformation of the protio and deuterio substrates have classically been determined in separate experiments, an experimental design that in and of itself introduces significant error. This is problematic because what really needs to be known to make sense of an isotope effect experiment and to be able to relate the magnitude of the observed isotope effect to the properties of the transition state and the mechanism of the reaction, is the intrinsic deuterium isotope effect for the reaction, i.e., the observed isotope effect needs to equal the intrinsic isotope effect. If the isotope effect is suppressed, for whatever reason, and the observed isotope effect is not equivalent to the intrinsic deuterium isotope effect, definitive conclusions regarding mechanism cannot be reached.

The experimental design that can best meet the criterion of an observed deuterium isotope effect being equivalent to an intrinsic deuterium isotope effect is an intramolecular

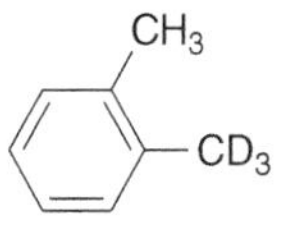

1,1,1-trideuteromethyl-*o*-xylene 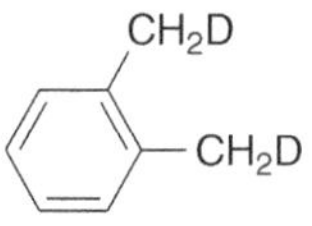   1,2-dideuteromethyl-*o*-xylene

FIGURE 3.7 Selectively deuterated *o*-xylenes for an intramolecular deuterium isotope effect experiment.

deuterium isotope effect experiment. In such an experiment, the structure of the substrate chosen has symmetry elements that allow the substrate to present the enzyme with an equal probability of operating on a carbon-deuterium bond or the equivalent carbon-hydrogen bond within the same molecule. This avoids the error associated with running two separate experiments and in addition normalizes all other potential partially rate-limiting steps so that only the bond-breaking step comes into focus. Under these conditions, the observed isotope effect is generally equivalent to the intrinsic isotope effect. For the experiment to work and yield the intrinsic isotope effect *what is absolutely necessary is that at all stages of reaction the active site of the enzyme sees equal concentrations of the carbon-hydrogen bond or the equivalent carbon-deuterium bond that is to be oxidized.* For example, in trying to determine the mechanism of cytochrome P450-catalyzed benzylic oxidation of xylene, a probe substrate that might be chosen to be used would be 1,1,1-trideuteromethyl-*o*-xylene (Fig. 3.7). When presented with this substrate, the enzyme would have the choice of oxidizing either the trideuteromethyl group or the equivalent normal methyl group adjacent to it. In order for equal concentrations of either the trideutero methyl group to be properly presented to the oxidative machinery of the active site of the enzyme the rate of interchange of the two groups within the active would have to be much faster than the actual rate of hydroxylation. Two adjacent methyl groups usually meet this criterion.

An even better substrate for examining the reaction would be 1,2-dideuteromethyl-*o*-xylene (Fig. 3.7). With this compound, the enzyme has the choice of oxidizing either a carbon-deuterium or a carbon-hydrogen bond at either methyl group. Motion of the substrate is not required and the rate of rotation of a methyl group can be assumed to be much faster than the rate of bond breaking. The requirement for the enzyme seeing equal concentrations of protium and deuterium during the course of the reaction (actually the enzyme sees two protiums for every deuterium), after statistical correction, is readily met.

For a more detailed and extensive exposition of the use of deuterium isotope effects, including other experimental designs, in drug metabolism studies the interested reader is referred to a recent review article and the references therein (7).

REFERENCES

1. Gibaldi M. Biopharmaceutics and Clinical Pharmacokinetics. Philadelphia, PA: Lea & Febiger; 1991.
2. Notari RE. Biopharmaceutics and Clinical Pharmacokinetics: An Introduction. New York: Marcel Dekker; 1987.
3. Silverman JA. P-glycoprotein. In: Levy RH, Thummel KE, Trager WF, et al., eds. Metabolic Drug Interactions. Philadelphia, PA: Lippincott, Williams & Wilkins; 2000.
4. Segel IH. Enzyme Kinetics: Behavior and Analysis of Rapid Equilbrium and Steady-State Enzyme Systems. New York: John Wiley & Sons, Inc.; 1993.

5. Neal JM, Kunze KL, Levy RH, et al. Kiiv, an in vivo parameter for predicting the magnitude of a drug interaction arising from competitive enzyme inhibition. Drug Metab Dispos 2003;31(8):1043–1048.
6. Levy RH, Trager WF. From in vitro to in vivo, an academic perspective. In: Levy RH, Thummel KE, Trager WF, et al., eds. Metabolic Drug Interactions. Philadelphia, PA: Lippincott, Williams & Wilkins; 2000.
7. Nelson SD, Trager WF. The use of deuterium isotope effects to probe the active site properties, mechanism of cytochrome P450-catalyzed reactions, and mechanisms of metabolically dependent toxicity. Drug Metab Dispos 2003;31(12):1481–1498.

4

Oxidation Pathways and the Enzymes That Mediate Them

For an organism to eliminate a lipophilic, chemically inert xenobiotic, it is usually first necessary to oxidize it to a more polar form. In addition, many biosynthetic pathways that produce steroid hormones, prostaglandins, leukotrienes, etc. involve oxidative steps. Organisms have evolved many enzymes to carry out these oxidations. Oxidation can occur by addition of oxygen (without addition of hydrogen which would represent hydration), removal of hydrogen atoms (without removal of oxygen which would represent dehydration), or simply removal of electrons.

OXIDATIVE ENZYMES

It is difficult to understand the oxidative pathways without an understanding of the enzymes that mediate them; therefore, we will start with a discussion of the oxidative enzymes with examples of specific oxidations that each enzyme meditates. This will be followed by a discussion of metabolic pathways organized according to functional groups.

Cytochrome P450s

Out of all the metabolic enzymes involved in the oxidation of drugs, the cytochrome P450s are by far the most common and the most important. They constitute a super family of membrane-bound enzymes whose individual members are found in virtually all living organisms from bacteria to the human. In humans, the P450s are found throughout the body but the highest concentrations are localized in the body's chemical factory—the liver. At the subcellular level, the P450s are found in the endoplasmic reticulum. Upon cellular homogenization followed by $100000 \times g$ centrifugation, the P450s present in endoplasmic reticulum can be isolated as a subcellular fraction known as microsomes. This is the primary enzymatic preparation that has been utilized for the last several decades for in vitro drug metabolism studies. These studies are now often complimented with experiments

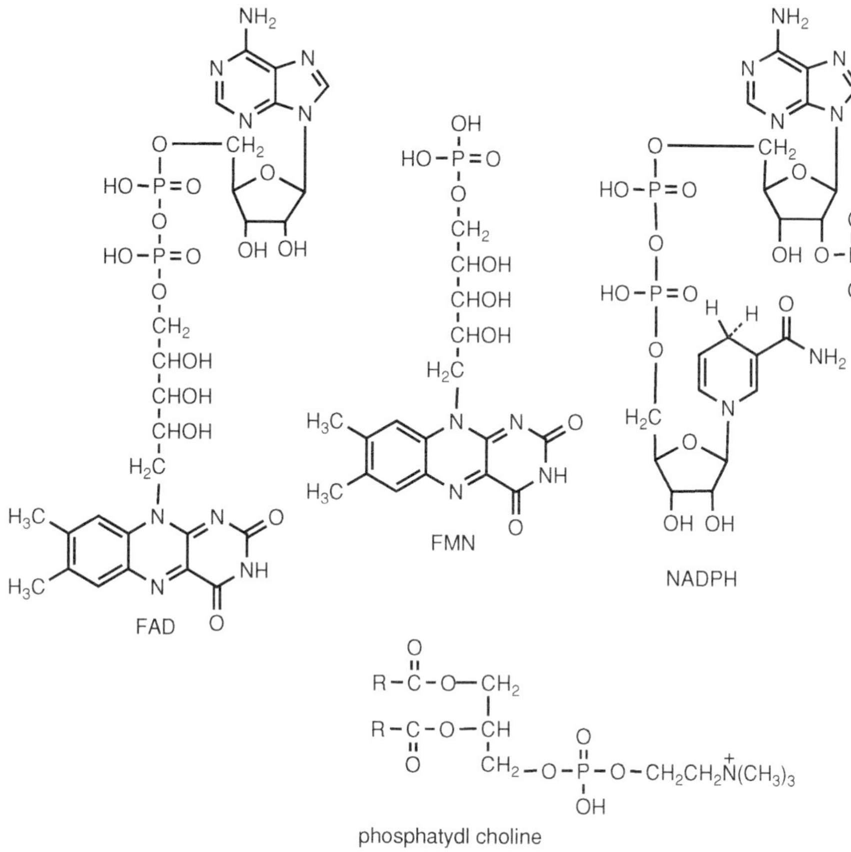

FIGURE 4.1   Other molecules required for P450 function.

using individual human P450s. These enzymes have become commercially available as a result of advances in molecular biology and have led to enormous advances in the field.

Cytochrome P450: General Properties and Mechanism of Oxygen Action
The P450s are moderately sized proteins having molecular weights that fall within the range of 48 to 53 kDa. The catalytic component of P450 is a heme cofactor, and the enzyme utilizes the redox chemistry of the Fe^{3+}/Fe^{2+} couple to activate molecular oxygen to oxidize and chemically modify drug molecules. The complete functional system also involves a second enzyme, cytochrome P450 reductase. Cytochrome P450 reductase is a 190-kDa protein that has both flavin adenine dinucleotide (FAD) and flavin adenine mononucleotide (FMN) as cofactors that serve to sequentially transfer reducing equivalents from reduced nicotine adenine dinucleotide phosphate (NADPH) to cytochrome P450 (Fig. 4.1). NADPH cannot reduce cytochrome P450 directly; the heme Fe^{3+} of P450 can only accept electrons in discrete single electron steps, whereas reduction by a hydride (H^-) ion from NADPH is a two-electron process. However, either the FMN or FAD cofactors of cytochrome P450 reductase can undergo a direct two-electron H^- reduction by NADPH and then transfer the electrons to P450 in single one-electron steps. In addition to cytochrome P450 and cytochrome P450 reductase, in vitro functional enzyme systems require a third component, lipid, which is generally supplied as phosphatidyl choline

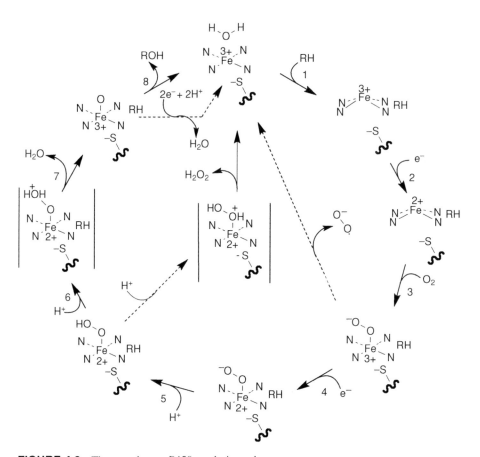

FIGURE 4.2 The cytochrome P450 catalytic cycle.

(Fig. 4.1). Since the P450s are membrane-bound enzymes in vivo, adding lipid to reconstitute a functional enzyme system in vitro more closely mirrors the in vivo environment and presumably serves as a matrix to allow the two enzymes, P450 and P450 reductase, to interact properly.

Cytochrome P450s work by activating molecular oxygen (O_2). They are all classified as mono-oxygenases because in the overall catalytic process, O_2 is split into two oxygen atoms but only one atom is utilized in oxidizing the substrate (RH) while the second atom is reduced by two electrons to form water [Eq. (4.1)].

$$CYP + O_2 + RH + 2e^- + 2H^+ \rightarrow CYP + ROH + H_2O \qquad (4.1)$$

The cytochrome P450 catalytic cycle (1) is shown in greater detail in Figure 4.2. The cofactor, heme, is anchored to P450 protein via an ionic interaction between the positively charged heme iron and a negatively charged cysteine thiolate residue from the protein. Heme iron is in the +3 oxidation state in the enzyme-resting state with the two remaining positive charges being counterbalanced by two negatively charged nitrogen atoms from the porphyrin ring. In the resting state, the electronic configuration of heme Fe^{3+} can exist in either low-spin or high-spin forms, but the low-spin form predominates. It is characterized by hexacoordinated heme Fe^{3+} in which Fe^{3+} lies in the plane of the porphyrin ring. Four of the ligand sites are occupied by the four imidazole nitrogens and the fifth by cysteine thiolate. The sixth is presumed to be occupied by a molecule of water.

Step 1. The substrate, RH, associates with the active site of the enzyme and perturbs the spin-state equilibrium. Water is ejected from the active site and the electronic configuration shifts to favor the high-spin form in which pentacoordinated heme Fe^{3+} becomes the dominant form-binding substrate. In this coordination state, Fe^{3+} is puckered out and above the plane in the direction of the sixth ligand site. The change in spin state alters the redox potential of the system so that the substrate-bound enzyme is now more easily reduced.

Step 2. NADPH-dependent P450 reductase transfers an electron to heme Fe^{3+}.

Step 3. O_2 binds, but can also dissociate. If it dissociates, the enzyme reverts to the heme Fe^{3+} resting state and generates superoxide radical anion in the process.

Step 4. A second electron, via P450 reductase or in some instances cytochrome b5, is added to the system generating a heme-bound peroxide dianion formally equivalent to FeO_2^+.

Step 5. H^+ adds to the system generating a heme-bound hydroperoxide anion complex formally equivalent to heme FeO_2H^{2+}.

Step 6. A second H^+ is added. If H^+ adds to the inner oxygen of heme, FeO_2H^{2+} decoupling occurs, H_2O_2 is released, and the enzyme reverts to the heme Fe^{3+} form.

Step 7. If the second H^+ adds to the outer oxygen of heme FeO_2H^{2+}, water is formed and released. Residual heme FeO^{3+} bears an oxygen atom (oxene) complexed to heme Fe^{3+}, a species considered to be analogous to compound 1, the reactive intermediate of the peroxidases. Decoupling can again occur via a two-electron reduction of FeO^{3+} plus the addition of two protons. This generates a molecule of water and the heme Fe^{3+} resting state of the enzyme. The degree to which this process occurs depends on the relative rates of heme FeO^{3+} reduction versus oxygen atom transfer to the substrate as outlined in the next step (2).

Step 8. An oxygen atom is transferred from heme FeO^{3+} to the substrate forming oxidized product, the product is released, and the enzyme reverts to its heme Fe^{3+} resting state.

Mechanism of Oxygen Atom Transfer . Oxene, the cytochrome P450 heme FeO^{3+} oxygen atom–bound catalytic species, is a highly reactive general oxidant. It will oxidize virtually any organic molecule with which it comes in contact including unactivated hydrocarbons (3). The energy required for catalysis is paid up front in generating heme FeO^{3+} so that P450s operate much differently than normal enzymes. Most enzymes operate by paying the energy price for reaction later in catalysis. Binding energy is used to orient substrate so that the conformation of bound substrate approaches the conformation of the transition state for reaction thus lowering both the entropic and enthalpic components of the energy of activation (1). This difference in operational properties accounts for why most enzymes are highly substrate selective while the P450s are highly promiscuous in terms of both the structure and chemical class of the substrate. Another major difference between most enzymes and the P450s is in the nature of their operational dependence on protein composition and active site architecture. For most enzymes, protein composition and active site architecture are intimately associated with the mechanism of reaction. In contrast, because of the high reactivity of heme FeO^{3+} coupled to its low substrate selectivity, protein composition and active site architecture appear not to be critical to the mechanism of P450 catalysis. Rather these elements play a dominant role in controlling substrate access to the active site oxidant through steric hindrance and/or ionic interactions. Because all P450s have the same heme cofactor and the same active oxidant, it is the spectrum of the specific substrates that are able to access the active site of any given P450 that distinguishes one P450 from another. This has at least three important ramifications:

1. A single P450 is capable of selectively oxidizing a substrate molecule at a number of different sites producing multiple metabolites. The number, identity, and relative importance of metabolites produced often reflect reaction at the energetically most easily oxidized substrate sites.
2. A single P450 is generally capable of oxidizing many different substrates within a chemical class as well as oxidizing substrates in a number of different chemical classes.
3. A number of different P450s often contribute to the production of the same metabolite from a given substrate.

Oxene Hydrogen Atom Abstraction . Because the active P450 oxidant, FeO^{3+}, is isoelectronic with carbene, i.e., oxene like carbene has only six electrons in its outer valence shell, its mode of reaction might be expected to be similar to that of carbene. Carbenes are known to react with carbon–hydrogen bonds by a direct insertion mechanism, i.e., a reaction in which the carbene inserts between a carbon–hydrogen bond in a single step. Such reactions are known to proceed with retention of configuration and are normally accompanied by a small deuterium isotope effect. However, cytochrome P450–catalyzed oxidation of covalent carbon–hydrogen bonds of simple normal hydrocarbons does not meet these expectations.

Early studies with a purified and reconstituted rabbit P450, CYP2B4 (earlier name P450 LM2) on the hydroxylation of the hydrocarbon, norbornane, found that the reaction proceeded to yield a 3.4:1 mixture of exo–endo norborneols (4). Further, hydroxylation of exo-2,3,5,6-tetradeuteronorbornane was found to proceed with a large deuterium isotope effect (11.5 ± 0.5) and a significant amount of epimerization in forming the endo and exo metabolites (Fig. 4.3).

The large isotope effect suggested that carbon–hydrogen bond cleavage occurs via a linear and symmetrical transition state, while the loss of stereochemical integrity via epimerization suggested the involvement of an intermediate. A mechanism that is consistent

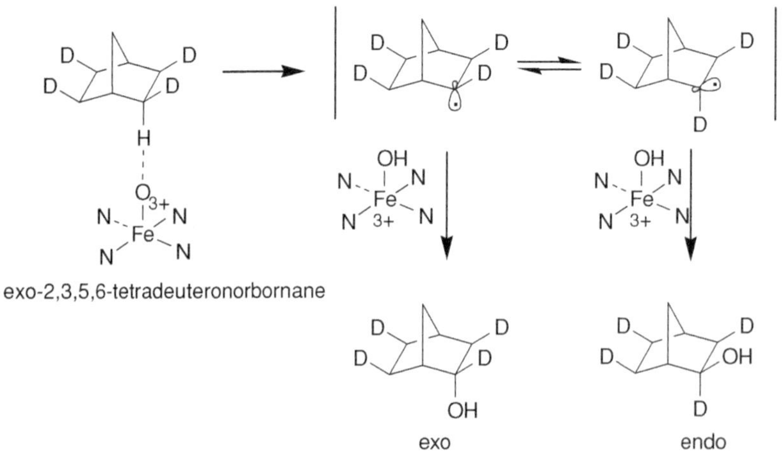

exo-2,3,5,6-tetradeuteronorbornane

exo   endo

FIGURE 4.3 P450-catalyzed oxidation of exo-2,3,5,6-tetradeuteronorbornane.

with these data is one in which the enzyme's heme-bound activated oxygen abstracts an endo hydrogen atom from exo-tetradeuteronorbornane to generate two radicals; a carbon-based radical that is sufficiently long lived to epimerize and a heme-stabilized hydroxyl radical. Subsequent combination of the hydroxyl radical with each of the two carbon radicals generated by epimerization, in what is termed the "oxygen rebound step," yields the isomeric alcohol products. Product is released and the ferric form–resting state of the enzyme is regenerated (Fig. 4.3). This radical rebound (hydrogen atom abstraction–oxygen rebound) mechanism has become the consensus mechanism for the cytochrome P450–catalyzed oxidation of covalent carbon–hydrogen bonds of simple normal hydrocarbons or hydrocarbon side chains in more complex molecules. It may also be the mechanism of heteroatom dealkylation, i.e., the oxidative cleavage of alkyl groups attached to heteroatoms (primarily N, O, and S) in drug molecules [see "Oxidation α to a Heteroatom (N, O, S, Halogen)" section in this chapter].

While the evidence for a radical rebound mechanism is strong, it is not without problems. When the rates of methyl group hydroxylation of a series of substituted methyl cyclopropanes were determined, serious anomalies emerged (5). In a study of the P450-catalyzed oxidation of a series of alkyl-substituted methylcyclopropanes, termed radical clocks because of the known lifetime of the carbinyl radicals generated chemically from each of these substrates, the rate of the oxygen rebound step was timed (6). If a radical rebound mechanism is assumed, then the rate of the oxygen rebound step can be determined from the ratio of un-rearranged products (cyclopropylmethanols) to rearranged products (alkenols) times the rate of known carbinyl radical rearrangement (rate of cyclopropyl ring opening). Thus, the ratios of *trans*-2-methylcyclopropylmethanol, 1, to the sum of the ring-opened alkene products, 3 and 4, and *cis*-2-methylcyclopropylmethanol, 2, to the sum of the same ring-opened alkene products, 3 and 4, formed from *trans*-1,2-dimethylcyclopropane and *cis*-1,2-dimethylcyclopropane, respectively, indicated that the average rate of the oxygen rebound step was $(1.7 \pm 0.2) \times 10^{10}$ per second (Fig. 4.4).

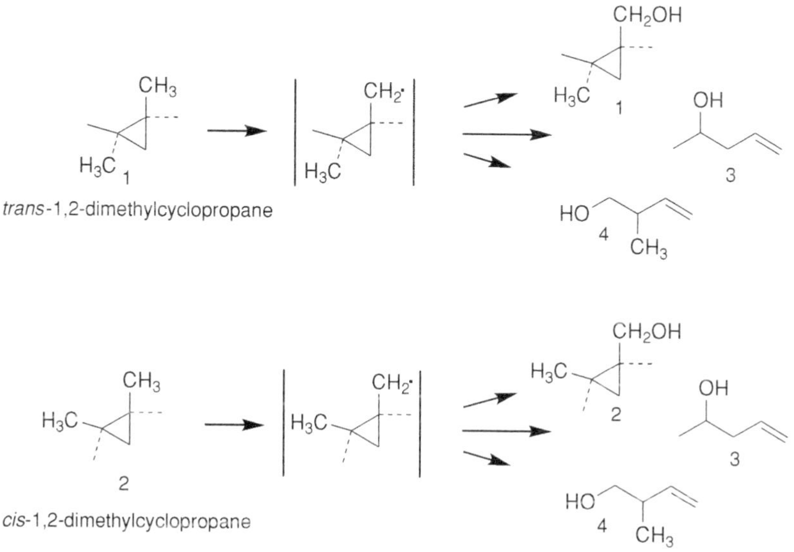

FIGURE 4.4 Cytochrome P450-catalyzed oxidation of *trans*-1,2-dimethylcyclopropane and *cis*-1,2-dimethylcyclopropane.

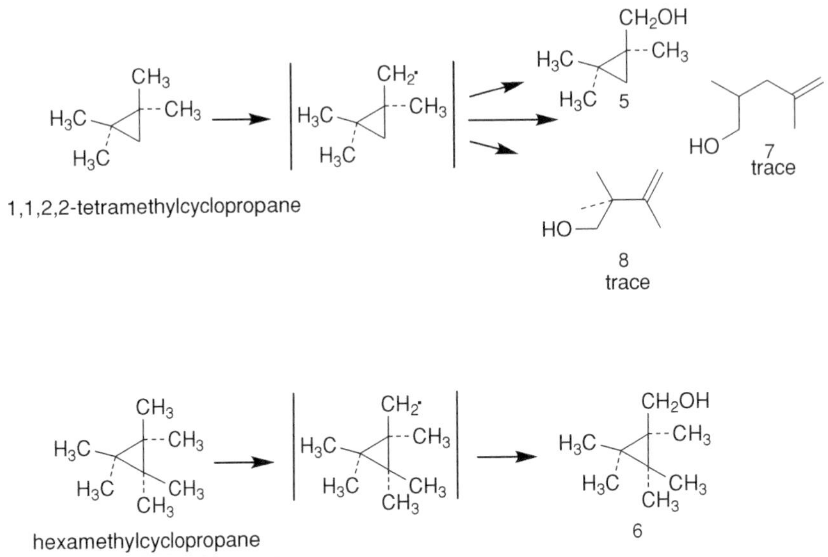

FIGURE 4.5 Cytochrome P450-catalyzed oxidation of 1,1,2,2-tetramethylcyclopropane and hexamethylcyclopropane.

Since the rates of rearrangement of carbinyl radicals formed from the sterically congested systems, 1,1,2,2-tetramethylcyclopropane and hexamethylcyclopropane, are much faster than those for *trans*-1,2-dimethycyclopropane and *cis*-1,2-dimethylcyclopropane, the relative amounts of rearranged products (to unrearranged products) from P450-catalyzed oxidation of 1,1,2,2-tetramethylcyclopropane and hexamethylcyclopropane were expected to be significantly greater than the relative amounts of rearranged products obtained from the P450-catalyzed oxidation of *trans*-1,2-dimethycyclopropane and *cis*-1,2-dimethylcyclopropane. This was found not to be the case as the hydroxylation of 1,1,2,2-tetramethylcyclopropane gave 1,2,2-trimethylcyclopropylmethanol, 5, and only traces of ring-opened alkene rearrangement products, 7 and 8, while hexamethylcyclopropane gave 1,2,2,3,3-pentamethylcyclopropylmethanol, 6, and a rearrangement product could not even be detected (Fig. 4.5) (6).

To address these seemingly paradoxical findings, the rat microsomal- and the CYP2B1-catalyzed hydroxylation of constrained methylcyclopropyl analog, 9, an ultrafast radical clock, was studied (7). This substrate yields both the hydroxymethylcyclopropyl product, 10, and a mixture of ring-opened rearrangement products, diastereomeric alcohols, 11 and 12 (Fig. 4.6). The rate of oxygen rebound required to be consistent with the observed reaction profile and the known rate of cyclopropyl ring opening was found to exceed 1.4×10^{13} sec^{-1}. However, the maximum theoretical rate allowable for the involvement of an intermediate is 6×10^{12} sec^{-1}. Thus the reaction rate is so fast that the lifetime of the intermediate carbinyl radical, 9', appears insufficient to be considered a true intermediate.

These results are inconsistent with a radical rebound mechanism because this mechanism is a two-step process that requires the involvement of intermediates. Instead the results suggest that the hydroxylation is a concerted process, much like a singlet carbene reaction, which does not involve intermediates. However, this conclusion is in conflict with the properties of singlet carbene reactions discussed above. Subsequent studies on a number of substituted methylcyclopropanes and other stained hydrocarbon systems established that these findings were not anomalous.

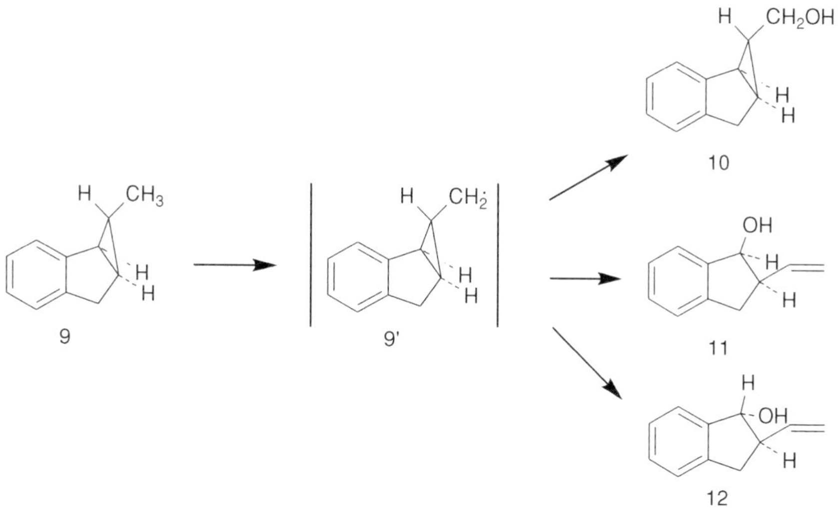

FIGURE 4.6   Cytochrome P450-catalyzed oxidation of constrained methylcyclopropane 9.

A potential resolution to the dilemma presented by results that are consistent with a two-step process, radical recombination mechanism (4) together with results that are consistent with a single step, direct insertion mechanism has recently been offered (8) based on theoretical calculations. A two-state reactivity paradigm that involves the interplay of two reactive states of the FeO^{3+} that are close in energy, a quartet state (high-spin state) and a doublet state (low-spin state), is invoked. In this description, the hydroxylation reaction proceeds in two distinct phases, an initial bond activation phase followed by a rebound phase. In the initial phase, FeO^{3+} attacks an alkyl hydrogen leading to the cleavage of the corresponding carbon–hydrogen bond and formation of a complex consisting of a heme-stabilized hydroxy radical and an alkyl carbon radical. In the subsequent rebound phase, the alkyl carbon radical reacts with the hydroxyl radical to form the carbon–oxygen bond of the alcohol product.

The transition states and ground states for reaction of the bond activation phase are similar in structure and close in energy (8), approximately 0.2 kcal/mole, for proceeding by either the high-spin or low-spin state forms of FeO^{3+}. In contrast, the rebound phase for the two-spin state forms is quite different. The high-spin state form reaction pathway proceeds in two steps. After the initial bond-breaking step to generate the heme-stabilized hydroxy radical and an alkyl carbon radical complex, the subsequent rebound step requires a new transition state with a significant barrier. The barrier to rebound is generally greater than the energy required to dissociate the complex to a heme-stabilized hydroxyl radical and a free alkyl radical. Thus, some dissociation is likely to occur (8). Formation of a free alkyl radical can lead to both epimerization and the formation of rearranged products. These results are consistent with the consensus radical recombination mechanism (4). The rebound step for the low-spin state form has no barrier, i.e., the energy required to reach the transition state for the rebound step is virtually zero. It proceeds to product formation in a single-step reaction pathway that involves a concerted but nonsynchronous mechanism (8). A concerted mechanism proceeds with retention of configuration, negating the possibility of epimerization. The lack of a free alkyl carbon radical also means that rearranged products will not be formed. These results are consistent with a single-step, direct insertion mechanism.

Since the two-spin state forms can lead to different products, the products obtained will be a mixture that reflects the initial fractionation of the reaction between the two-spin states. The fractionation in turn is a reflection of the interplay and the probability of cross-over between the two-spin states (8). Thus, the two-state reactivity paradigm resolves the dilemma of whether a radical recombination or a direct insertion mechanism governs cytochrome P450–catalyzed hydroxylation; actually they are both involved and the degree to which either is expressed depends upon the specific substrate hydroxylated and the specific enzyme.

Cytochrome P450 Nomenclature
All P450s carry a CYP prefix indicating that they are the members of the P450 superfamily. The CYP prefix is followed by an Arabic numeral, 1, 2, 3, . . ., etc. indicating the specific family to which the individual enzyme belongs. Family members share at least a 40% sequence identity. The family designation is followed by a capitalized letter of the alphabet, A, B, C, . . ., etc. that designates the subfamily to which the individual enzyme belongs. Subfamily members share at least a 55% sequence identity. Finally, the individual enzyme is identified by an Arabic numeral following the alphabet letter. A typical cytochrome P450 enzyme might have the designation, CYP3A4, which would indicate it was the fourth individual cytochrome P450 identified as belonging to the A subfamily of the 3 family. The specific gene that encodes for a specific P450, e.g., CYP3A4, would carry the same designation except it would be italicized, i.e., *CYP3A4*.

The specific P450s involved in human drug metabolism are found primarily in three families, CYP1, CYP2, and CYP3. Out of the various individual members of these families, CYP1A2, CYP2A6, CYP2B6, CYP2C9, CYP2C19, CYP2D6, CYP2E1, and CYP3A4 appear to be the major contributors. Of these CYP3A4 is probably the most important human enzyme as it has been found to be a significant contributor to the metabolism of approximately half of all drugs in current medical use (9). After consideration of the mechanism of action and the general properties of the P450s, each of the isoforms listed above will be highlighted in the following discussion.

Individual Human Cytochrome P450s . As indicated earlier, members of three families of cytochrome P450, CYP1, CYP2, and CYP3, dominate human drug metabolism and the primary property that distinguishes one P450 from another is the difference in the spectrum of activity displayed by each individual isoform in their ability to discriminate between substrates. This is largely by controlling the active site access through differing steric and/or electronic interactions. This being the case, it is informative to consider an overview of the properties of each of the major human forms of P450 in order to anticipate favored substrate types for each.

CYP1. The CYP1A subfamily contains two members, CYP1A1 and CYP1A2, which are involved in drug metabolism and have sparked considerable interest because they also seem to be associated with the metabolic activation of procarcinogens to mutagenic species.

CYP1A1. In humans, of the two members, CYP1A2 is the major player while CYP1A1 is a relatively minor extrahepatic isoform associated with the oxidation of polycyclic aromatic hydrocarbons like benzo[a]pyrene. Similarly, in test rodent species it is responsible for the generation of toxic intermediates and carcinogenic metabolites (10).

CYP1A2. CYP1A2 has been implicated in the activation of procarcinogenic species such as aflatoxin B1, 2-acetylaminofluorene, and other arylamines. It tends to favor aromatic substrates, both heterocyclic aromatic substrates like caffeine and aromatic substrates like phenacetin (10). In the case of caffeine, 1A2 is the major isoform catalyzing the

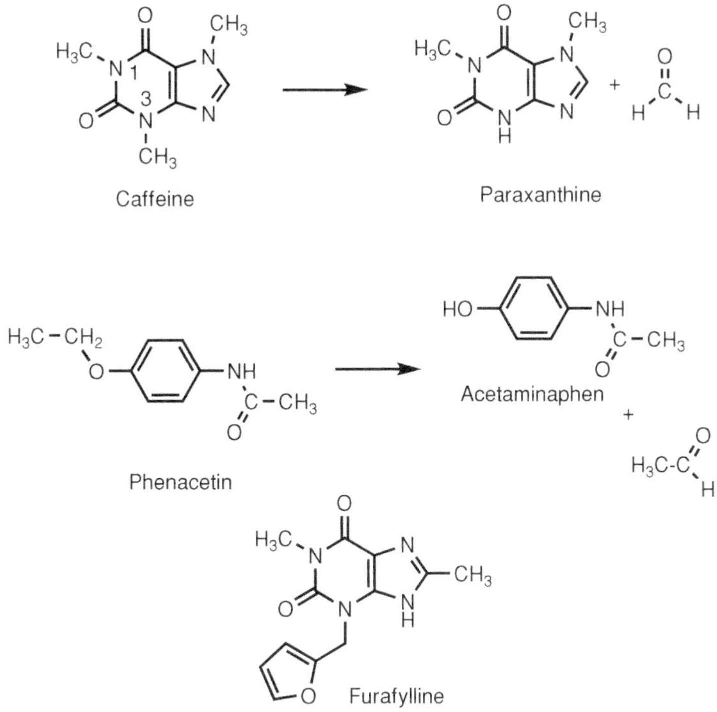

FIGURE 4.7   Structures of the CYP1A2 substrates caffeine, phenacetin, their metabolites, and the CYP1A2 mechanism–based inhibitor, furafylline.

N-demethylation at the three N-methyl sites. In this regard, the 3-N-demethylation of caffeine to generate paraxanthine can serve as a particularly good in vivo indicator of the presence and activity of CYP1A2 (Fig. 4.7). In the case of phenacetin, CYP1A2 catalyzes N-deethylation to generate acetaminophen. Not unexpectedly, 1A2's selectivity toward heterocyclic aromatic substrates carries over to inhibitors of the enzyme. Furafylline (Fig. 4.7) is an example of a particularly potent 1A2 mechanism-based inhibitor.

CYP2. The CYP2 family contains isoforms from at least five subfamilies, 2A, 2B, 2C, 2D, and 2E, which contribute significantly to the drug metabolism. Recognized members of each subfamily are enumerated and a brief description of the major contributor from each follows.

CYP2A6. The 7-hydroxylation of coumarin (11) and the initial carbon hydroxylation of the α-carbon to the pyrrolidine nitrogen of nicotine, which upon further oxidation by aldehyde oxidase (AO) (discussed later in this chapter) yields cotinine, are the defining metabolic activities associated with CYP2A6 (Fig. 4.8). CYP2A6 is also responsible for the stereospecific 3'-hydroxylation of cotinine to form *trans*-3'-hydroxycotinine (12), a major metabolite of nicotine in the human. CYP2A6 is polymorphic and its activity has a significant effect on smoking behavior. People with reduced or deficient CYP2A6 levels demonstrate a significantly reduced dependency upon nicotine (13). While CYP2A6 is the primary P450 responsible for nicotine metabolism, only a few other substrates have thus far been identified where CYP2A6 serves a similar role (14). The few that have been identified suggest that the active site of CYP2A6 favors small aromatic or heteroaromatic substrates, alkoxy ethers, and N-nitrosoalkylamines that are neutral or basic in character.

CYP2B6. While generally accounting for significantly less then 1% of the total P450 present in human liver, CYP2B6 is also found in extrahepatic tissue, including

coumarin   7-hydroxycoumarin

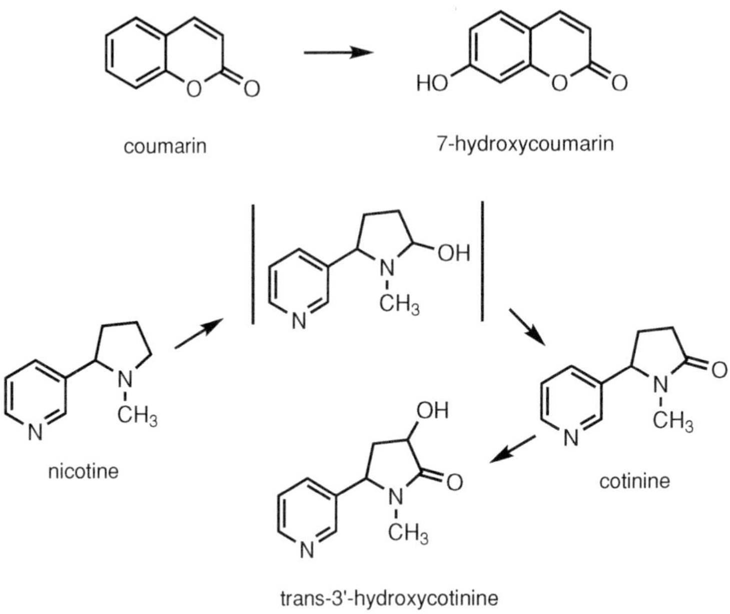

nicotine

cotinine

trans-3'-hydroxycotinine

FIGURE 4.8   Structures of the CYP2A6 substrates, coumarin and nicotine, and their metabolites.

brain, and it has been established as a major catalyst for the oxidation of several important drugs in current clinical use. For example, CYP2B6 catalyzes the 4-hydroxylation
and the N-dechlorethylation of the anticancer agents cyclophosphamide and ifosfamide
(15), respectively, the 4-hydroxylation of the anesthetic agent propofol (16), and the
methyl group hydroxylation of the antidepressant and antismoking agent bupropion (17)
(Fig. 4.9).

The O-deethylation of 7-ethoxy-4-trifluoromethylcoumarin has been the favored substrate to probe for CYP2B6 activity (18), but recent evidence indicates that it is not as selective for CYP2B6 as one would hope because both CYP1A2 and CYP2E1 also catalyze this
reaction. A much better indicator of CYP2B6 activity appears to be the N-demethylation
of (S)-mephenytoin, particularly at higher concentrations of (S)-mephenytoin (Fig. 4.10).
With the development of more selective CYP2B6 indicator substrates, the spectrum of
CYP2B6 catalytic activity will become more clearly defined as will its contribution to the
metabolism of major human medications.

CYP2C9. CYP2C9 is the most abundant isoform of the CYP2C subfamily (CYP2C8,
CYP2C9, CYP2C18, and CYP2C19) and one of the most extensively characterized of
all the human P450s. The active site has been explored with a variety of substrates, and
computer-derived homology models that predict substrate affinity have been developed.
The enzyme displays a distinct preference for acidic substrates with the defining substrates
being warfarin, tolbutamide, and the nonsteroidal anti-inflammatory drugs (NSAIDS).
Typical examples of the later are flurbiprofen and diclofenac (Fig. 4.11).

In the case of warfarin, CYP2C9 stereoselectively catalyzes the 7-hydroxylation
and 6-hydroxylation of (S)-warfarin to generate both (S)-7-hydroxywarfarin and (S)-6-
hydroxywarfarin in a ratio of 3:1. Together the two biologically inactive metabolites account
for better than 80% of the clearance of (S)-warfarin from the body (19). Since (S)-warfarin
is responsible for most of the drugs anticoagulant activity [(S)-warfarin is five to eight
times more potent an anticoagulant than (R)-warfarin], CYP2C9 effectively controls the
level of anticoagulation by controlling the in vivo concentration of (S)-warfarin, a drug with

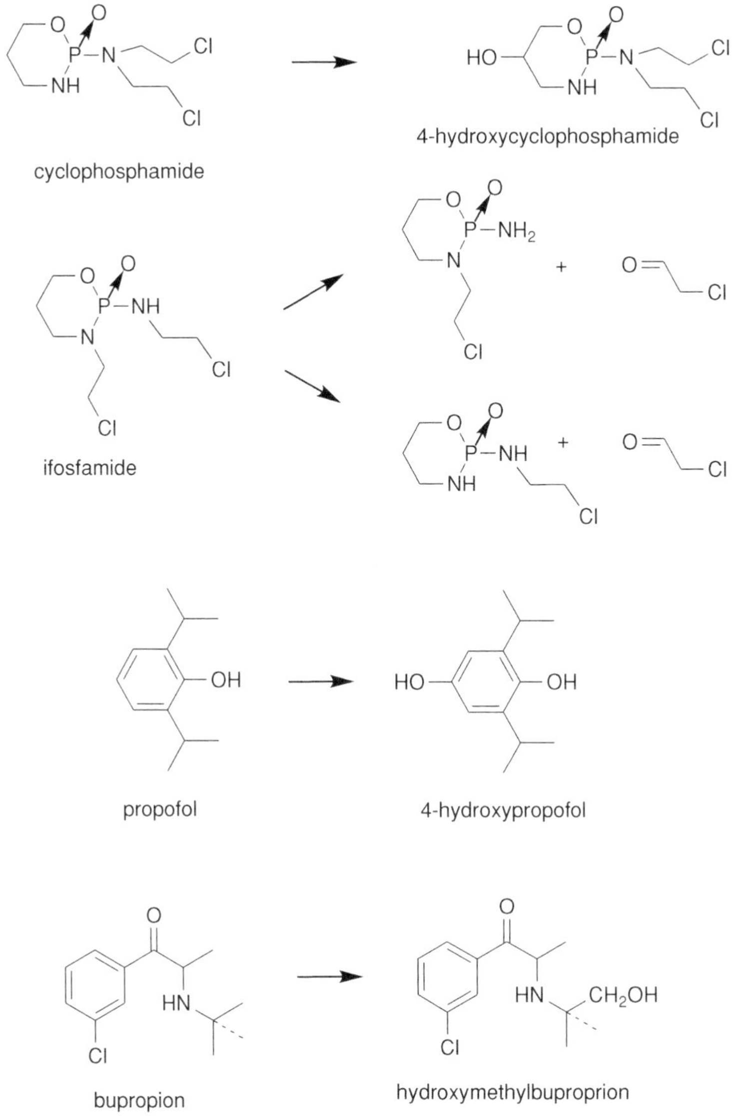

cyclophosphamide

4-hydroxycyclophosphamide

ifosfamide

propofol

4-hydroxypropofol

bupropion

hydroxymethylbuproprion

FIGURE 4.9 Structures of the CYP2B6 substrates, cyclophosphamide, ifosfamide, propofol, bupropion, and their metabolites.

a narrow therapeutic index. As a consequence, interference with CYP2C9 activity could be expected to have a major impact on anticoagulant response. Thus, if a second drug, in addition to warfarin, were present in vivo, and if the second drug were either a substrate and/or inhibitor of CYP2C9, a serious drug interaction could result. This indeed seems to be the case as a number of warfarin drug interactions have been shown to be caused by a second drug inhibiting CYP2C9 (20), and the metabolic inactivation of (*S*)-warfarin as a direct consequence. In this regard, it is informative to note that while (*R*)-warfarin is not a substrate of CYP2C9 it is a reasonably potent inhibitor ($K_i = 8\ \mu$M) of the enzyme and does affect the elimination rate of (*S*)-warfarin ($K_m = 4\ \mu$M) when the drug is administered as a racemate, its normal mode of administration (21). Therefore, while the two enantiomers of the drug have comparable affinities for the enzyme, one enantiomer is a substrate while the

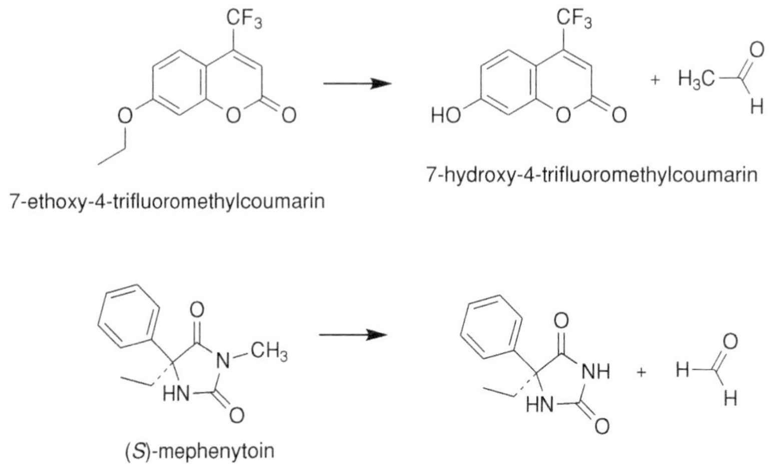

FIGURE 4.10 Structures of the CYP2B6 substrates, 7-ethoxy-4-trifluoromethylcoumarin and (*S*)-mephenytoin, and their metabolites.

other is an inhibitor. Clearly, interactions of this type help define the nature of the active site and form the basis for the development of substrate prediction.

Hydroxylation of the benzylic methyl group of tolbutamide, the preferred site of oxidative attack by CYP2C9 (22), generates hydroxytolbutamide. Hydroxytolbutamide is rapidly oxidized by other enzymes, presumably aldehyde oxidase and/or alcohol dehydrogenase (ALD), to form the major isolated metabolite, the benzoic acid analog.

The major CYP2C9-catalyzed transformation of (*S*)-flurbiprofen is formation of (*S*)-4′-hydroxyflurbiprofen (23) and that of diclofenac is formation of 4′-hydroxydiclofenac (24). A recently determined crystal structure of flurbiprofen-bound CYP2C9 indicates that the interaction of the carboxylate anion of flurbiprofen with a complex of hydrogen-bonded residues, Arg-108, Asp-293, and Asn-289, orients the substrate for regioselective hydroxylation (25). Moreover, the identification of this anionic-binding site helps explain how CYP2C9, an enzyme that has a relatively large active site, is able to catalyze the regioselective hydroxylation of small molecules such as the NSAIDS with high catalytic efficiency.

CYP2C19. While CYP2C19 is not a major human P450, it does illustrate two features of this enzyme family that are worth highlighting. First, it is 91% structurally homologous with CYP2C9 and yet the two enzymes have distinct substrate selectivities (26). It is not particularly active in metabolizing the substrates that characterize CYP2C9 nor does it favor anionic substrates. Defining substrates include the anticonvulsant, mephenytoin, and the proton-pump inhibitor, omeprazole, neither of which is a substrate for CYP2C9 (Fig. 4.12). This suggests that relatively limited structural changes can have profound effects on substrate selectivity despite the fact that all the P450s utilize the same activated oxygen species. Indeed, a change as limited as a single amino acid in an enzyme that is comprised of as many as 500 amino acids can have a major effect. For example, the I359L allelic variant of wild-type CYP2C9 is much less effective in metabolizing (*S*)-warfarin, the pharmacologically active enantiomer of racemic warfarin. In vitro kinetic analysis of CYP2C9 I359L indicated that the mutant metabolized (*S*)-warfarin with a fivefold lower V_{max} and a fivefold higher K_m than the wild-type CYP2C9 (27) suggesting that individuals who carried this mutant would be much more sensitive to the effects of the anticoagulant and require a much lower dose. This indeed has been found to be the case (28).

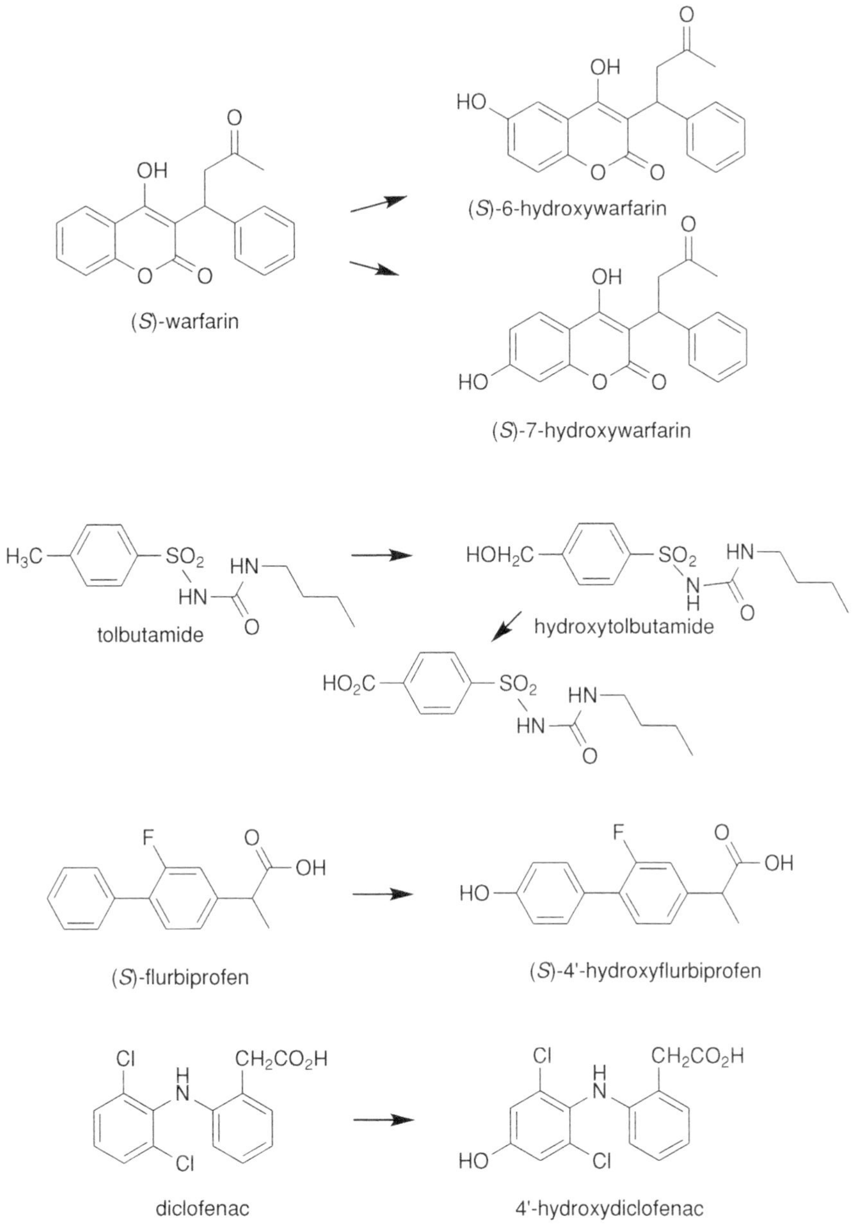

FIGURE 4.11 Structures of the CYP2C9 substrates, warfarin, tolbutamide, (S)-flurbiprofen, and diclofenac, and their metabolites.

The second important feature of CYP2C19 is that it is the first CYP to illustrate the potential importance of mutant forms of the enzyme to therapeutic outcome using standard dosing. The 4'-hydroxylation of (S)-mephenytoin is the major metabolic pathway leading to the elimination and termination of the anticonvulsant activity of (S)-mephenytoin. CYP2C19 is the cytochrome P450 that catalyzes this metabolic transformation. However, in early studies the ability to metabolize mephenytoin seemed to vary within the population such that two distinct groups could be identified: extensive metabolizers and poor metabolizers. It turns out that a defective mutant form of CYP2C19 is carried by 4% of Caucasians

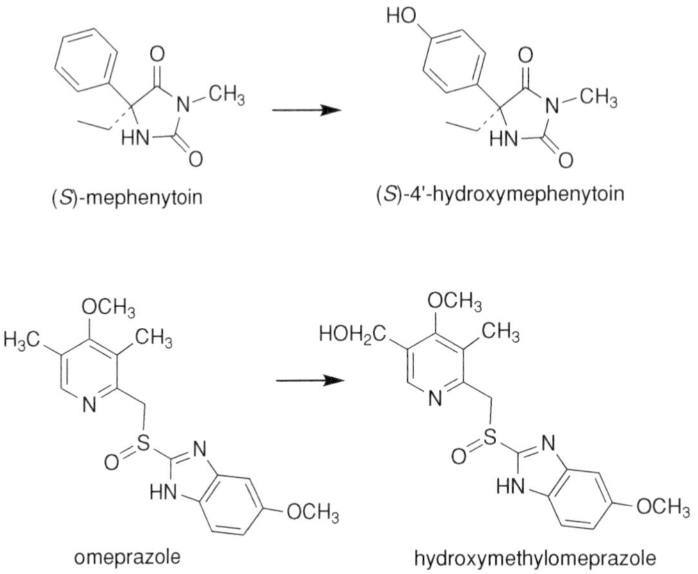

(S)-mephenytoin   (S)-4'-hydroxymephenytoin

omeprazole   hydroxymethylomeprazole

FIGURE 4.12 Structures of the CYP2C19 substrates, (S)-mephenytoin and omeprazole, and their metabolites.

but a full 20% of Asians. Thus, it is clear that if effective therapeutics is to be achieved, particularly with drugs with a narrow therapeutic index, knowledge of the metabolism of the drug and the enzymes and possible enzyme variants that control its metabolism is critical.

CYP2D6. While the amount of CYP2D6 present in human liver is generally less than 10% (9) of the total amount of cytochrome P450 present in human liver, it nevertheless can be considered a major contributor to the metabolism of a significant number of potent drugs used in clinical pharmacology. From the perspective of substrate preference, CYP2D6 can be considered the mirror image of CYP2C9. Where CYP2C9 has a preference for acidic substrates, CYP2D6 prefers basic substrates. Since most active central nervous system drugs are bases, it is hardly surprising that CYP2D6 plays an important role in the metabolic processing of these agents. A recent compilation listed 56 drugs where CYP2D6 is the primary or one of the major cytochrome P450s responsible for their metabolism (29). Typical examples include the benzylic hydroxylation of the antidepressive agent, amitriptyline, the O-demethylation of the analgesic, codeine, the N-dealkylation of the antipsychotic, haloperidol, and the 4-hydroxylation of the antihypertensive, propranolol (Fig. 4.13).

In contrast, the antiarrhythmic agent, quinidine (also a base), is a potent (sub-μM) inhibitor of the enzyme (Fig. 4.14). This fact illustrates that while the basic properties of quinidine insure that it has affinity for CYP2D6, affinity does not guarantee that the substrate will properly orient in the active site of the enzyme with respect to the active oxidant, FeO^{3+}, for efficient metabolic transformation. Thus quinidine could be considered as a " silent substrate" of CYP2D6, i.e., a compound that is a highly effective inhibitor by virtue of its affinity for the enzyme but one that is a poor substrate by its failure to achieve an efficient catalytically susceptible orientation. Silent substrates are potentially important causes of drug interactions because if they are present in vivo with another drug whose metabolism is governed by an enzyme that they potently inhibit, an exaggerated pharmacological response would result. Moreover, the exaggerated response would be

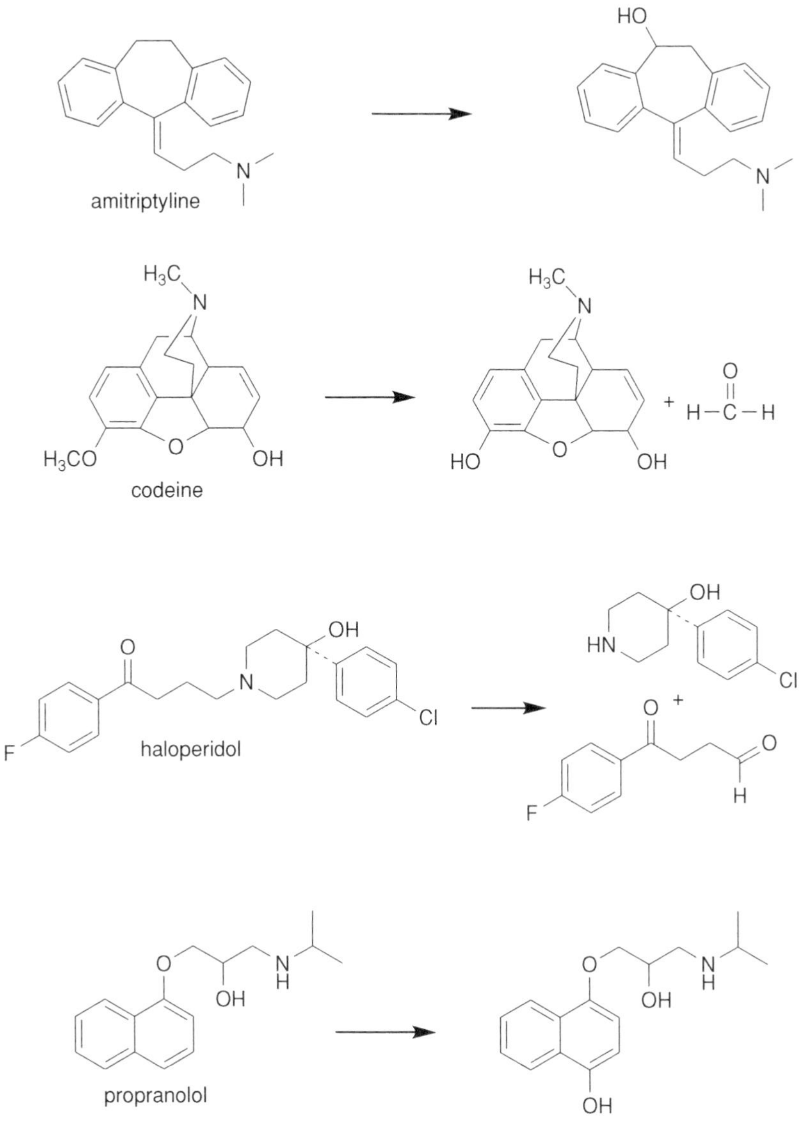

FIGURE 4.13   Structures of the CYP2D6 substrates, amitriptyline, codeine, haloperidol, and propranolol, and their metabolites.

totally unexpected unless the silent substrate had been prescreened for its ability to inhibit that particular enzyme.

The fact that so many substrates for CYP2D6 have the common structural feature of being organic bases is probably the reason that this particular cytochrome P450 was among the first human P450s whose active site was explored using the computer technique of homology modeling. While the crystal structures of mammalian P450s were totally unknown, the structure of a soluble bacterial P450, CYP101 (P450cam), had been solved and the solution of a few others, e.g., CYP102 (P450BM3) soon followed. Given that the amino acid sequence of CYP2D6 was known, the assumption that the tertiary structure of CYP101 would be preserved in CYP2D6 together with sequence-related residues allowed a computer-derived working model of the active site of CYP2D6 to be developed (30).

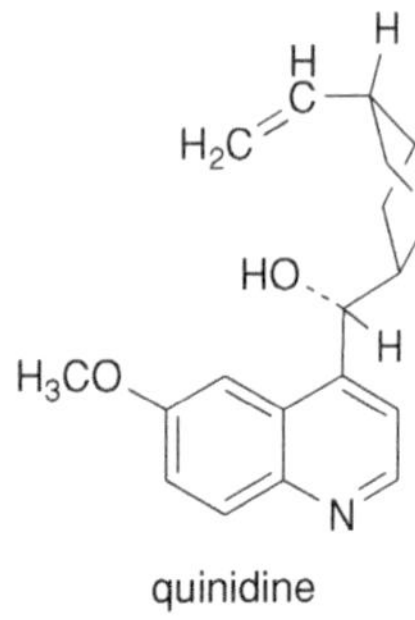

FIGURE 4.14 Structure of the CYP2D6 competitive inhibitor, quinidine.

This model suggested that Asp-301 was an active site residue that was critical in binding the substrate in the active site of the enzyme through ionic interaction with the protonated amino group of the substrate. Subsequent studies to refine the active site have supported the role of Asp-301 as well as indicating the role of other contributing residues (Phe-120, Thr-309, and Glu-216) (31,32). The recent publication of the crystal structure of CYP2D6 validates these catalytic assignments (33). Moreover, the study importantly confirms the initial assumption of the preservation of tertiary structure between mammalian and bacterial forms of cytochrome P450.

Like CYP2C19, CYP2D6 exhibits a common genetic polymorphism. In fact it was the first cytochrome P450 for which a genetic polymorphism was clearly established (34). Historically, the two drugs that defined the polymorphism and indicated that individuals within the European population could be categorized as either extensive metabolizers or poor metabolizers were the antihypertensive agent, debrisoquine, and the labor-inducing agent, sparteine (Fig. 4.15). About 5–10% of this population is found to be poor metabolizers and has little capacity to convert either of these two drugs to their major metabolites, 4-hydroxydebrisoquine and 5-dehydrosparteine. While the clinical usefulness of both the drugs has been superseded by the development of better agents, they can still be effectively used as analytical tools to evaluate the catalytic activity of CYP2D6 in vivo in an individual or in vitro in a liver sample.

CYP2E1. Chronic exposure of rats to ethanol leads to enhanced cytochrome P450 activity. After discovery of the phenomenon, the enhanced activity was soon characterized

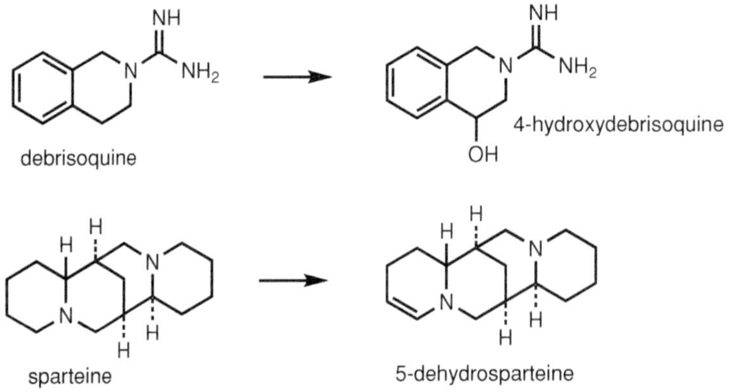

FIGURE 4.15 Structures of the CYP2D6 substrates, debrisoquine and sparteine, and their metabolites.

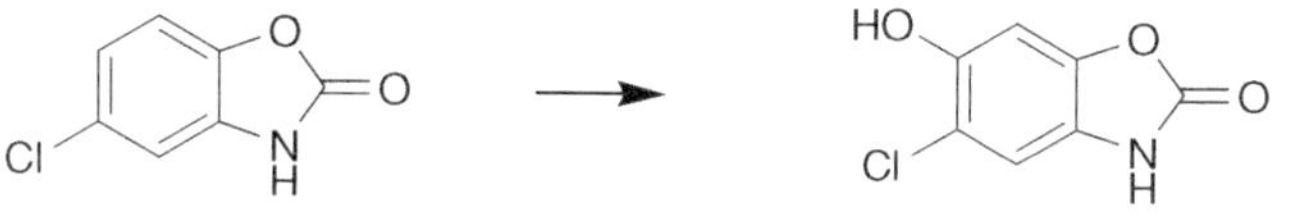

FIGURE 4.16 Structure of the CYP2E1 substrate, chlorzoxazone, and its metabolite, 6-hydroxychlorzoxazone.

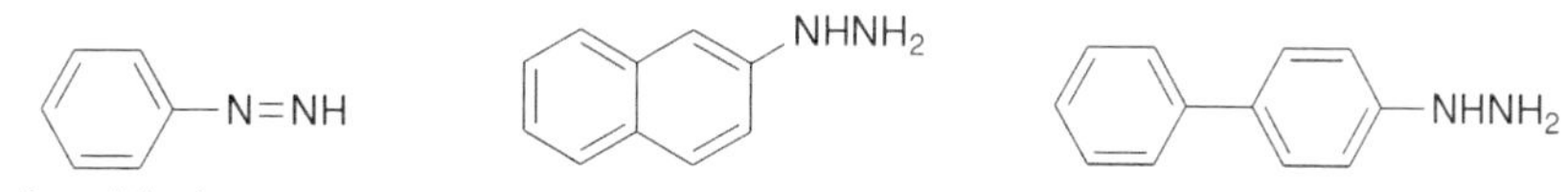

FIGURE 4.17 Structures of the CYP2E1 mechanism-based inhibitors, phenyldiazine, 2-naphthylhydrazine, and p-biphenylhydrazine.

as being primarily due to the induction of a single P450. This enzyme was subsequently identified as CYP2E1 (35). Still later studies with the selective CYP2E1 substrate, chlorzoxazone, confirmed that chronic ethanol ingestion also led to the selective induction of CYP2E1 in humans (Fig. 4.16).

Ethanol is both an inducer and substrate of CYP2E1. Indeed, CYP2E1 seems to be structurally geared to favor small volatile molecules such as ketones, aldehydes, alcohols, halogenated alkenes, and alkanes as substrates (36). Moreover, many of these same compounds, like ethanol, are inducers of the enzyme. A major mechanism by which this diverse group of compounds appears to initiate induction is by inhibiting normal enzyme degradation.

The apparent preference for small molecules suggests that CYP2E1 has a restricted active site. This simple observation is supported by the formation of aryl–iron complexes (Fe–Ar) in the reactions of human CYP2E1 with phenyldiazene, 2-naphthylhydrazine and p-biphenylhydrazine (Fig. 4.17) (37). The results indicate that the active site cavity is relatively open above pyrrole rings A and D but is closed above pyrrole rings B and C. These results are also supported by the results of a subsequent computer-generated homology model based on the coordinates of the soluble bacterial P450, cytochrome P450BM3 crystal structure, and analysis of the amino acid sequence of P4502E1 (38).

Since a number of CYP2E1 substrates are industrial chemicals to which large numbers of people are exposed, induction has significant toxicological implications. It turns out that the structural properties of many CYP2E1 substrates can lead to the formation of chemically reactive metabolites upon enzyme-catalyzed oxidation. It also turns out that a number of these reactive metabolites are either carcinogenic or generate the expression of other toxicities. For example, chloroform is converted to phosgene, other halohydrocarbons can similarly be metabolized to acid chlorides or reductively transformed to reactive radicals, e.g., CCl_4 to $\cdot CCl_3$, ethanol is converted to acetaldehyde, alkenes are converted to epoxides, e.g., butadiene to butadiene monoepoxide. In addition, CYP2E1 generates methyl carbonium ion, a reactive methylating species capable of methylating DNA, subsequent to the N-demethylation of tobacco-generated nitrosoamines, e.g., N,N-dimethylnitrosoamine, to N-methylnitrosoamine, to methyl carbonium ion, water, and nitrogen, (Fig. 4.18).

CYP3A4. Out of all the cytochrome P450s involved in human drug metabolism, CYP3A4 could be considered to be the most important by virtue of the fact that at least 50% of marketed drugs that are metabolized by P450s are metabolized by the CYP3A4 (9). Generally, it is the most abundant P450 present in human liver, averaging 29% in a study

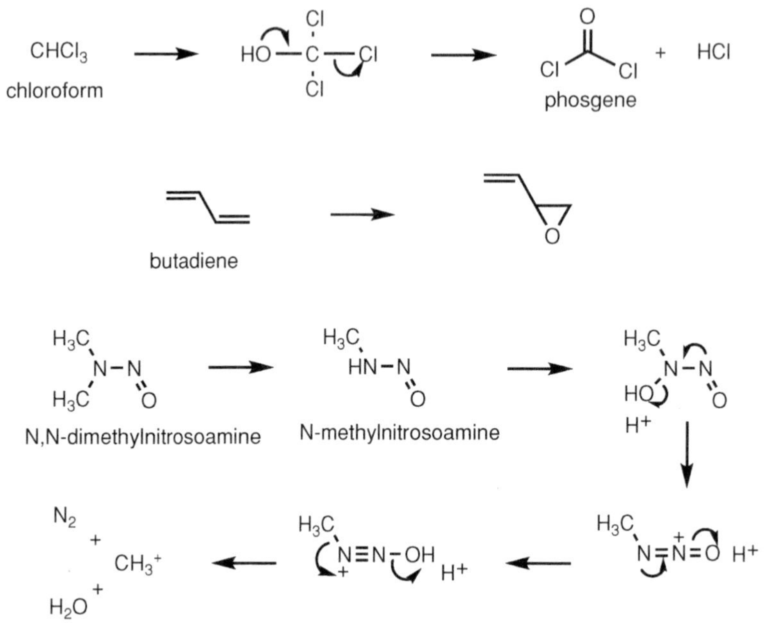

FIGURE 4.18 Structures of the CYP2E1 substrates, chloroform, butadiene, and *N,N*-dimethylnitrosoamine, and their chemically reactive and toxic metabolites.

that determined P450 content in 60 human liver samples. Like all P450s, percent content of any specific P450 can vary between individuals. The variability of CYP3A4 between individuals can be as high as 20-fold (39).

In addition to being the most abundant P450 in human liver, it is also the most abundant P450 in human intestinal mucosa averaging about 40% of what is found in liver. The high intestinal content of CYP3A4 can have a major effect on the bioavailability of orally administered drugs because any orally administered drug must first pass through the intestinal mucosa before reaching the systemic circulation. Thus, a significant fraction of a first-pass effect might be due to the passage through the intestine and exposure to CYP3A4 before the drug reaches the liver via the portal vein, where it is again exposed to metabolism before it enters the systemic circulation.

The broad range of structural types that are substrates for CYP3A4, in addition to the fact that many are relatively large molecules, e.g., macrocyclic antibiotics, suggests that the enzyme might have a large, relatively open (accessible) active site. Consistent with this indication is the fact that the intramolecular sites of hydroxylation of CYP3A4 substrates correspond to the sites that would be expected to be hydroxylated based on energetics, i.e., the sites based on physical organic chemistry are the energetically most susceptible sites to oxidation. This suggests that the substrate has freedom of motion within the active site so that it can achieve the orientation that allows the most energetically favored site to be oxidatively attacked.

The crystal structure of CYP3A4 has been solved both as unliganded enzyme and as enzyme bound to the inhibitor, metyrapone, or bound to the substrate, progesterone (Fig. 4.19) (40). In contrast to what might have been expected, the structures revealed a surprisingly small active site with little conformational change associated with the binding of either compound. This apparent anomaly could be explained if the active site, as might be expected, has greater freedom of motion in its natural biological environment than it

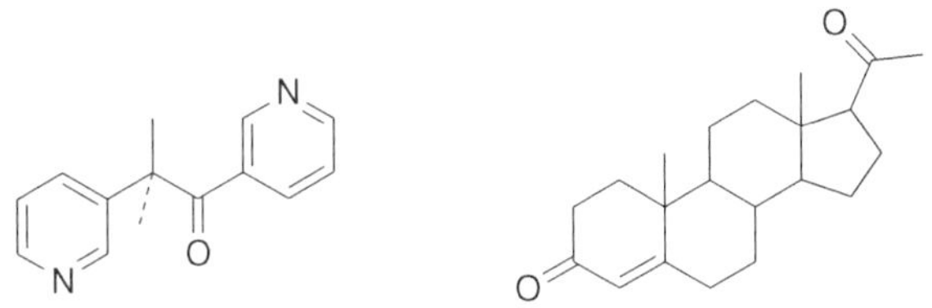

FIGURE 4.19 Structures of the CYP3A4 substrates, metyrapone and progesterone.

does in the solid crystalline state. The active site could "breathe," i.e., open and close to allow the entrance of larger substrates and then their release after catalytic transformation. An unexpected peripheral binding site was identified that may be involved as an allosteric effector site that could conceivably modulate such a process. A number of substrates of CYP3A4 have been used as in vivo and/or in vitro markers of the enzyme activity to determine CYP3A4 content in human subjects or in liver or intestinal preparations. Examples include the N-demethylation of erythromycin, the ring oxidation of nifedipine, the 6β-hydroxylation of testosterone, and the 1'-hydroxylation of midazolam (Fig. 4.20). Out of these examples, the 1'-hydroxylation of midazolam has properties that make it the method of choice, particularly as an in vivo probe. Midazolam is completely adsorbed, has a half-life of 60 to 90 minutes, the 1'-hydroxylation process is specific to CYP3A4 at the concentrations used, and it appears not to be a substrate for p-glycoprotein, the efflux pump present in the intestinal mucosa. This means that, if desired, it would be possible to independently assess the CYP3A4 content in liver and intestine within a subject by simultaneously administering oral and intravenous doses (one dose being labeled with a stable or radioactive isotope to distinguish it from the other dose) of midazolam.

    Since every substrate of the enzyme is also an inhibitor, the implication of the effectiveness of CYP3A4 in catalyzing the biotransformation of so many drugs in current use implies that at least potentially clinically significant drug interactions might be associated with the use of these drugs. Clearly, one might expect to observe a drug interaction when a drug primarily metabolized by CYP3A4 is co-administered with another medication that is also either a substrate or inhibitor of this enzyme. In clinical practice, however, this turns out not to be the major problem that might have been expected. In order for a significant interaction to occur, the enzyme must be substantially inhibited and this generally requires a concentration of the inhibitor at the active site of the enzyme well in excess of its K_i. For many inhibitors, the in vivo concentration achieved at the active site of the enzyme is less than its binding constant, i.e., K_i. Thus, significant interactions generally arise from very potent competitive inhibitors, i.e., ones with a K_i in the low micromolar or sub-micromolar region, or time-dependent inhibitors, i.e., ones that covalently modify the enzyme, e.g., MI complexes or suicide substrates.

Peroxidases

Peroxidases are heme iron-containing proteins similar in structure to that of cytochromes P450. The major difference is that peroxidases have histidine as the axial ligand instead of cysteine, and there are also other polar amino acids close to the heme iron that help to catalyze the peroxidase function of the enzyme (41). The result is that the peroxidases very rapidly catalyze the reduction of hydroperoxides to alcohols (or water in the case of

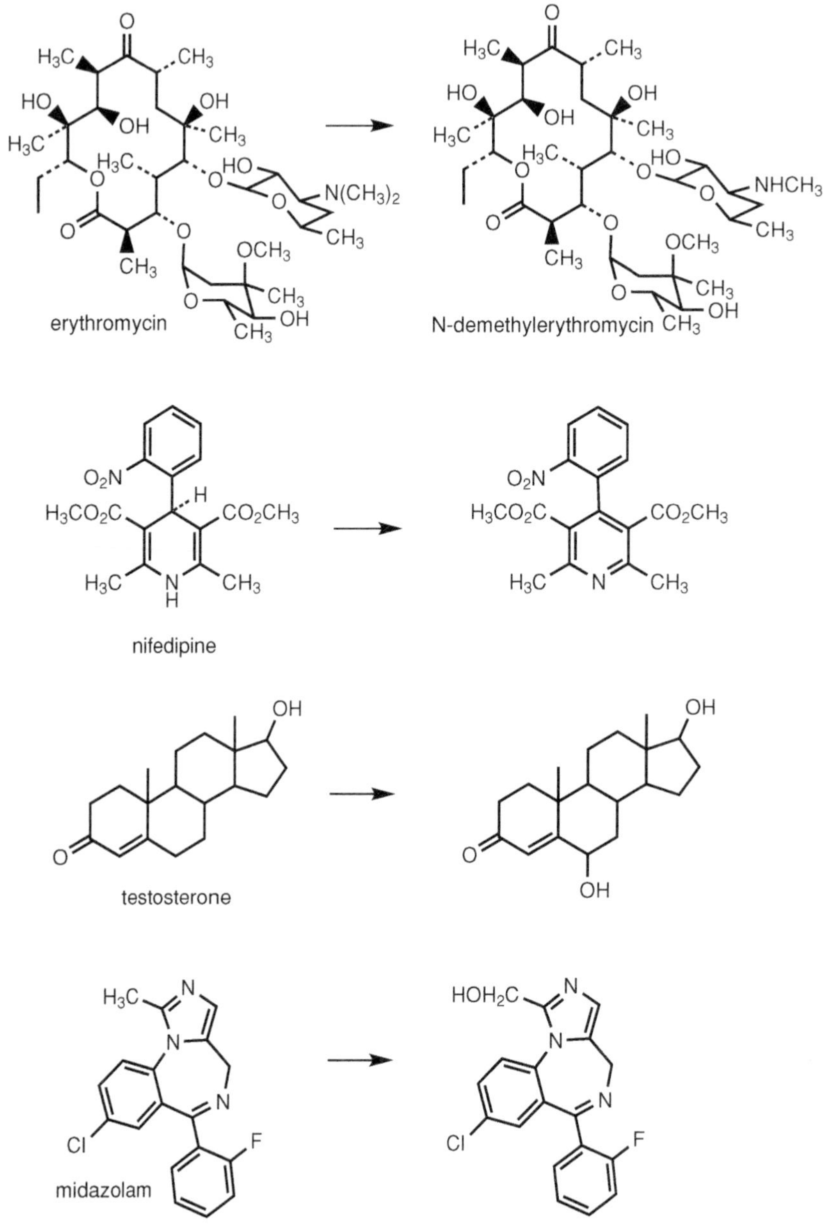

FIGURE 4.20 Structures of the CYP3A4 substrates, erythromycin, nifedipine, testosterone, and midazolam, and their metabolites.

hydrogen peroxide) with the concomitant oxidation of the peroxidase. Thus the catalytic cycle of peroxidases is simpler than that of the P450s as shown in Figure 4.21.

The oxidation of peroxidases by hydroperoxide leads to a ferryl iron–oxo species as well as a radical cation on the porphyrin ring, which is sometimes transferred to an adjacent amino acid. This species is referred to as compound I. Compound I can oxidize substrates directly by a two-electron process to regenerate the native peroxidase, but, more commonly, it oxidizes substrates by an one-electron process to form compound II where the porphyrin radical cation has been reduced. Compound II, in turn, can perform a second one-electron

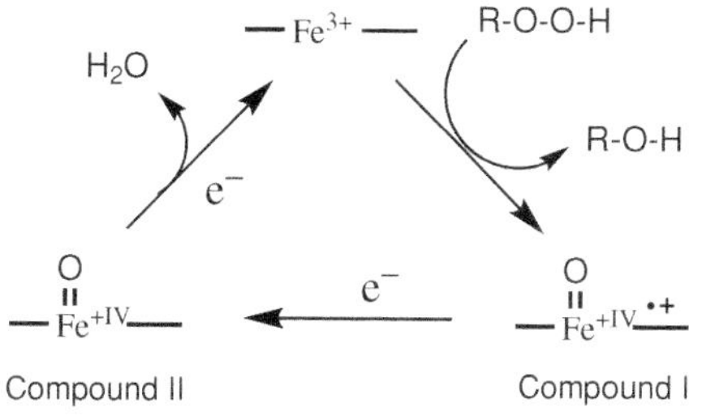

FIGURE 4.21 Catalytic cycle of peroxidases.

oxidation to generate the native enzyme. The second step is usually slower than the first reduction and thus is the rate-limiting step in the cycle. Unlike the cytochromes P450, which are capable of oxidizing almost anything, oxidation of xenobiotics by peroxidases is limited to electron-rich molecules that are easily oxidized.

Horseradish Peroxidase

Horseradish peroxidase, as the name implies, is derived from a plant not from humans or animals; however, it is readily available and often used as a model to study peroxidase oxidations (42). The classic substrates are phenols, which are oxidized to phenoxy radicals, but aromatic amines are also good substrates.

Prostaglandin H Synthase

Prostaglandin H synthase (also known as cyclooxygenase) utilizes two molecules of oxygen to oxidize arachidonic acid to prostaglandin G (43). Prostaglandin G is both an endoperoxide and a hydroperoxide (Fig. 4.22). This same enzyme also mediates the conversion of prostaglandin G to prostaglandin H, which involves reduction of the hydroperoxide, and in the process xenobiotics can be oxidized. This enzyme is widely distributed in mammals, and therefore it has the potential to oxidize substrates in locations where there is very little P450.

Prostaglandin synthase is present in the bladder and can oxidize aromatic amine carcinogens to reactive metabolites that bind to DNA, and people who are exposed to aromatic amines have an increased incidence of bladder cancer (44). What makes the bladder a target for this carcinogenicity is that the aromatic amines or hydroxylamines are often glucuronidated and excreted in the urine where they reach significant concentrations. Human urine is acidic and the low pH hydrolyzes the glucuronide back to the aromatic amine or hydroxylamine. This, in turn, can be oxidized to the ultimate carcinogen by prostaglandin H synthase.

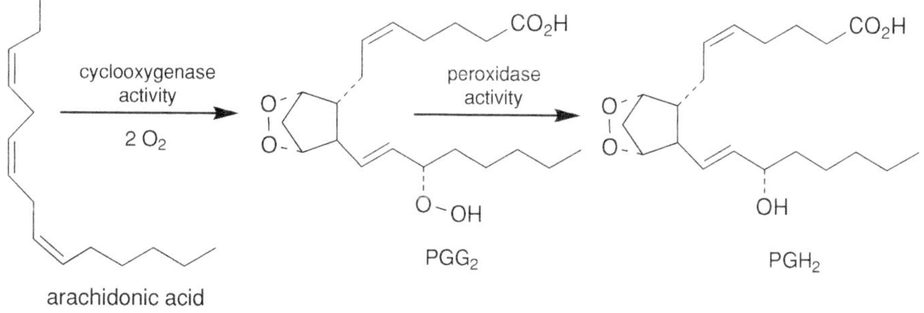

FIGURE 4.22 Prostaglandin H synthase–mediated conversion of arachidonic acid to PGH_2.

$$H^+ + {}^-O-Cl \rightleftharpoons H-O-Cl \xrightarrow{Cl^-} Cl-Cl + {}^-O-H$$

FIGURE 4.23 The oxidation potential of hypochlorite is dependent on pH and Cl^- concentration.

Another location with high prostaglandin-synthase activity and low P450 levels is the fetus, and it has been proposed that the teratogenicities of benzo[*a*]pyrene and phenytoin are due to cyclooxygenase-mediated bioactivation of these agents (45,46).

Prostaglandin synthase is also present in the skin where it may make a significant contribution to the oxidation of carcinogens and the pathogenesis of skin cancer (47).

Myeloperoxidase

A major difference between myeloperoxidase and most other peroxidases is that the substrate is chloride ion, which is oxidized to hypochlorous acid, and in the process compound I is converted directly to the native peroxidase instead of going through compound II (48). Myeloperoxidase is found in neutrophils, and when these cells are activated they have another enzyme, NADPH oxidase, which generates hydrogen peroxide (NADPH oxidase actually generates superoxide that is converted to hydrogen peroxide). Hypochlorite is very effective in killing bacteria and viruses and that is a major function of neutrophils. Hypochlorite is also added to most drinking water to kill bacteria and viruses, and this is based on the oxidant properties of hypochlorous acid. These oxidations are pH dependent because hypochlorous acid has a pK_a of 7.5 (49); therefore, in vivo approximately half of it is present as ClO^-, which is not a strong oxidant and does not contribute to the oxidant properties of HOCl. In the presence of excess chloride ion, hypochlorous acid is in equilibrium with Cl_2, which is an even stronger oxidant than HOCl. Given the relative oxidation potentials of hypochlorite, hypochlorous acid, and molecular chlorine, it can be seen from Figure 4.23 that the oxidizing potential of hypochlorous acid is increased by a low pH and high chloride concentration.

HOCl can also oxidize drugs. There are several drugs that can cause agranulocytosis (an absence of granulocytes which are mostly neutrophils). In general, these drugs are also oxidized to reactive intermediates by HOCl. Therefore, it is likely that the reactive intermediates formed by HOCl are responsible for causing agranulocytosis. Examples of HOCl-generated reactive intermediates of drugs associated with agranulocytosis are clozapine (50), amodiaquine (51), aminopyrine (52), and vesnarinone (53) as shown in Figure 4.24.

Clozapine and vesnarinone are discussed in Chapter 8. It is interesting to note that while other peroxidases, such as horseradish peroxidase and prostaglandin H synthase, oxidize aminopyrine by one-electron oxidation to a relatively stable blue radical cation (54), oxidation by myeloperoxidase appears to be a two-electron oxidation that generates a very reactive dication (52). Monocytes also contain myeloperoxidase and can oxidize drugs. It is possible that the ability of monocytes to form reactive metabolites and their role as precursors to antigen-presenting cells may play an important role in the induction of other idiosyncratic drug reactions, especially drug-induced lupus (55).

Even though neutrophils generate HOCl and, in general, the same products are produced by myeloperoxidase and HOCl, it appears that in some cases the oxidant is a chlorinated form of the myeloperoxidase rather than HOCl (55). It might be expected that little oxidation of such drugs would occur in vivo because it would require activation of the neutrophils in order to generate hydrogen peroxide; however, using an antibody that recognizes clozapine bound to protein, it was found that neutrophils from patients who take clozapine have substantial amounts of clozapine bound to their neutrophils (56).

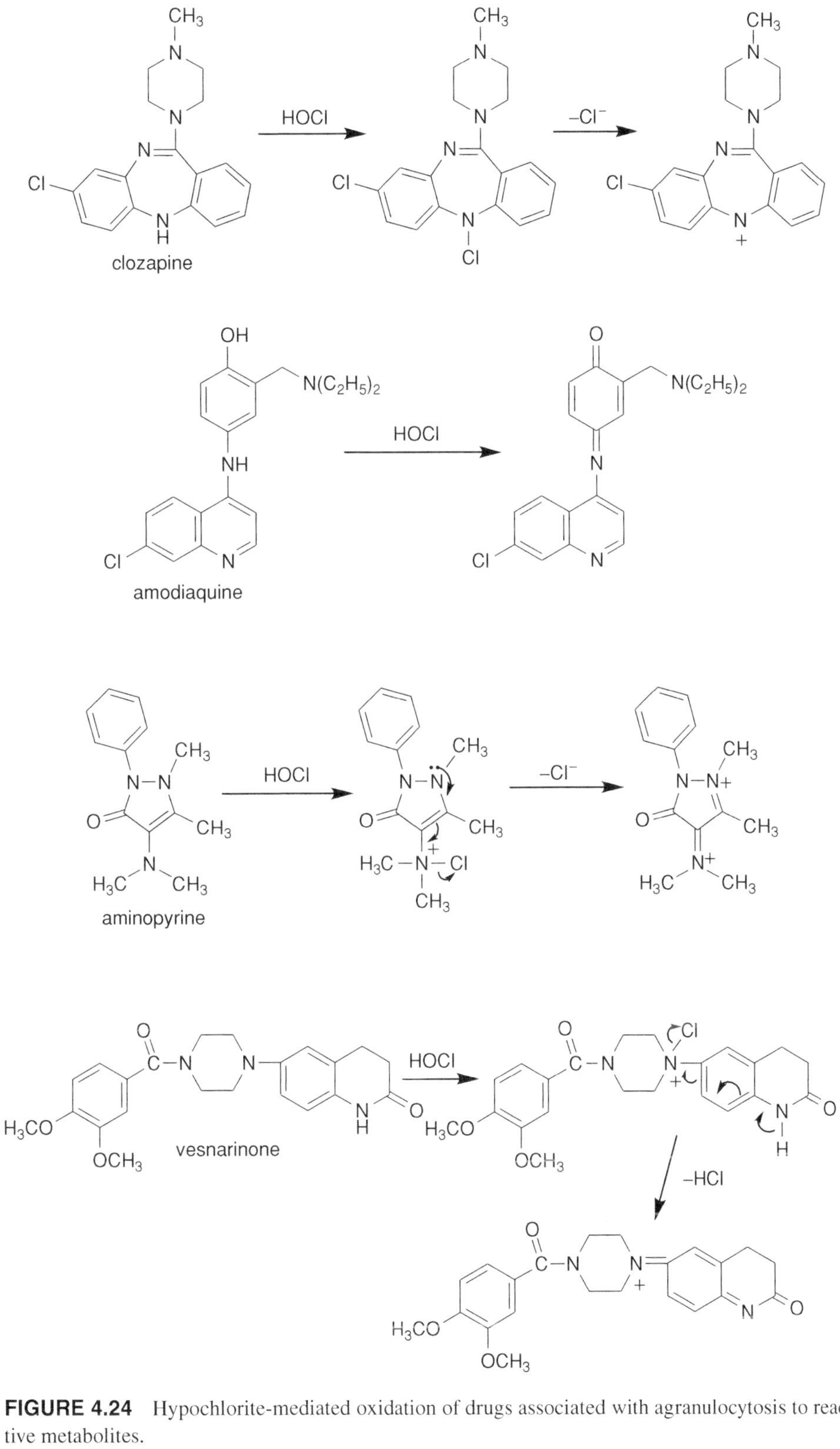

FIGURE 4.24 Hypochlorite-mediated oxidation of drugs associated with agranulocytosis to reactive metabolites.

Miscellaneous Peroxidases

As the name implies, lactoperoxidase peroxidase is produced in the breast and is found in breast milk. It has been proposed that oxidation of carcinogenic amines by lactoperoxidase may contribute to breast cancer (57,58). Eosinophil peroxidase is similar to myeloperoxidase except that chloride is not a good substrate; bromide and thiocyanate are much better substrates for this enzyme (59). Eosinophils appear to be most important for combating parasites, but a genetic lack of eosinophil peroxidase is generally clinically silent. Thyroid peroxidase utilizes iodide for the synthesis of thyroxin in the thyroid gland. It is also able to oxidize some drugs, and drugs such as minocycline are oxidized in the thyroid gland resulting in a dark pigmentation of the gland and inhibition of thyroid function (60).

Flavin Monooxygenase

The flavin monooxygenases (FMOs) are widely distributed in nature and have multiple members with six (FMO1, . . . FMO6) having been characterized (61). In humans, the most abundant form is FMO3 and, like P450, is found in the cellular endoplasmic of the liver. Interestingly, FMO1, the form of the enzyme that occurs in fetal liver, relocates to the kidney in adults while only a trace remains in the liver (62). Like the P450s, the FMOs display broad substrate selectivity but are largely complimentary to the P450s in terms of substrate selectivity. Nature has designed the FMOs to oxidize a class of compounds that the P450s are much less efficient in oxidizing: foreign compounds that contain electron-rich polarizable nucleophilic groups. These are largely the compounds containing the elements of sulfur, nitrogen, selenium, and phosphorus, e.g., thiols, sulfides, disulfides, 1°, 2°, and 3° amines, imines, hydrazines, hydroxylamines, selenols, selenides, phosphines, etc. (63).

The mammalian FMOs operate by the two-electron oxidation characteristic of peroxides rather than by the radical-like sequential one-electron oxidations characteristic of the P450s. The active oxidizing species of the FMOs is the C(4a)-hydroperoxyflavin (Fig. 4.25). It is generated from the catalytic cycle that is initiated by the NADPH binding to the enzyme and reducing the isoalloxazine ring of FAD to the dihydro form, $FADH_2$, which is

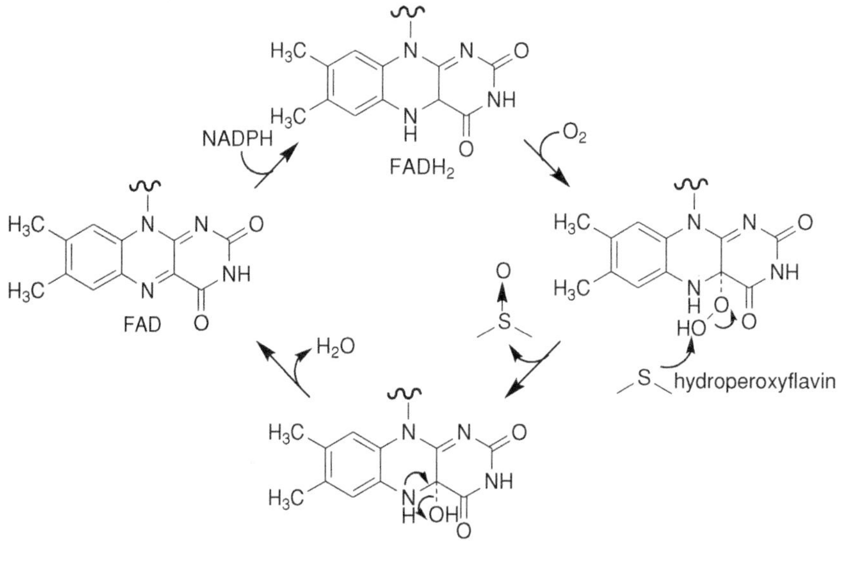

FIGURE 4.25 FMO catalytic cycle.

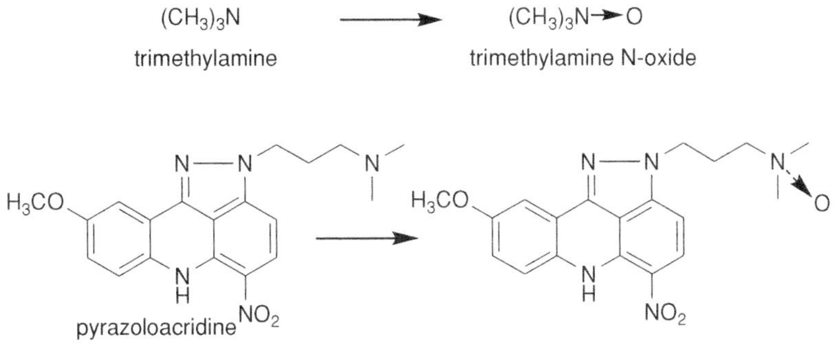

FIGURE 4.26   Structures of the FMO substrates, trimethylamine and pyrazoloacridine, and their N-oxide metabolites.

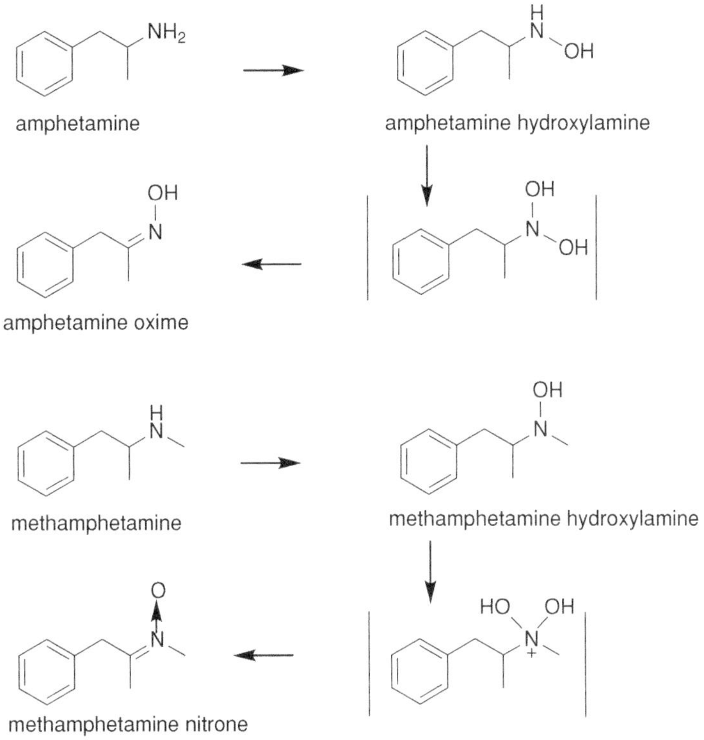

FIGURE 4.27   FMO-catalyzed oxidation of amphetamine and methamphetamine and the structures of their metabolic products.

susceptible to oxidation. Oxygen adds to the enzyme and oxidizes the reduced isoalloxazine ring to form the active hydroperoxyflavin-oxidizing species. Substrate is oxidized and water is eliminated from the residual C(4a)-hydroxyflavin to regenerate FAD and complete the catalytic cycle (64,65). The reactivity of the active oxidizing species accounts for the broad substrate selectivity of these enzymes and suggests that, like the P450s, a primary criterion for reaction is substrate access to the oxidizing species (66).

     FMOs and P450s are found in the same intracellular site and often catalyze the same reaction with a given substrate but with different efficiencies. Distinguishing which enzyme is responsible for a reaction or how much a particular enzyme contributes in the case

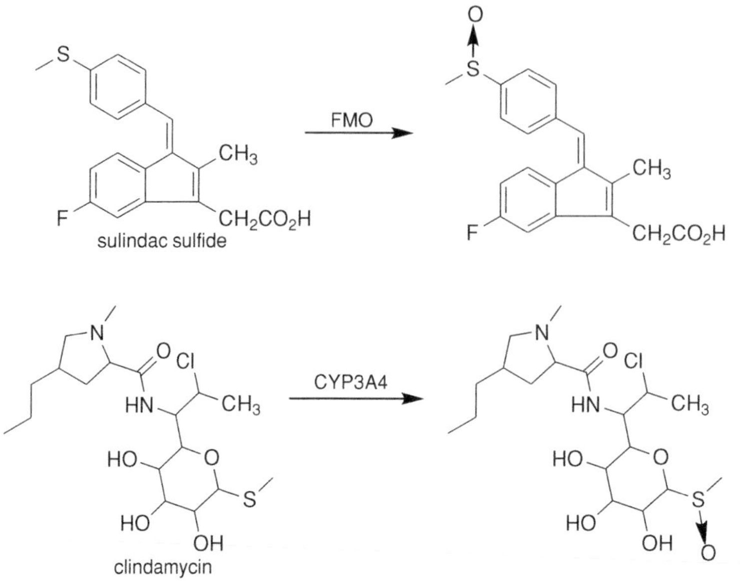

FIGURE 4.28 Oxidation of sulindac sulfide to sulindac sulfoxide by FMO and oxidation of clindamycin to clindamycin sulfoxide by CYP3A4.

where more than one enzyme contributes to the reaction becomes particularly important in in vitro–in vivo correlation and drug interaction studies. Fortunately, catalytic activities due to either FMO or P450 can readily be distinguished (67) one from the other, simply by heat or running microsomal reactions at higher pH. If the microsomal preparation is heated (45°C for 5 minutes) before NADPH is added to initiate the reaction, P450 maintains its activity while that of FMO is lost. If in a separate experiment the pH of the incubation mixture is raised to 9, FMO maintains activity but P450 activity is virtually abolished, particularly in the presence of detergent.

One of the better-known reactions of the FMOs is the oxidation of tertiary amines to form N-oxides. One particularly important reaction in this regard from a physiological point of view is the FMO3-catalyzed oxidation of dietary-derived trimethylamine to the odorless trimethylamine N-oxide (68). Individuals with trimethylaminuria can reek of this foul-smelling amine as an unfortunate consequence of having a diminished capacity to oxidize it. The diminished capacity to oxidize the amine has been shown to be due to the mutations in the gene encoding for FMO3. A recent example of a drug being converted to its N-oxide by FMO can be found in the metabolism of the experimental pyrazoloacridine antitumor agent (69) (Fig. 4.26).

FMO also oxidizes primary and secondary amines. For example, it N-hydroxylates both amphetamine and methamphetamine to generate the corresponding hydroxylamines (Fig. 4.27) (70). It then catalyzes a second N-hydroxylation of both metabolites. The two N,N-dihydroxy intermediates eliminate water to generate the oxime in the case of amphetamine and the nitrone in the case of methamphetamine.

While N-oxidation is essentially driven by FMO, it is not true of S-oxidation. As indicated earlier, P450 can contribute significantly to S-oxidation and, in some cases, it is the dominant or even only enzyme catalyzing the reaction. Sulindac sulfide, a metabolite of the nonsteroidal anti-inflammatory agent, sulindac, is reoxidized by FMO (71) with a high degree of stereoselectivity toward the (R)-enantiomer back to enantiomerically

TABLE 4.1 Kinetic Parameters for the Oxidation of
Alcohols by Alcohol Dehydrogenase

Alcohol	Km (mM)	Vmax (relative to ethanol)
Methanol	7.00	0.09
Ethanol	0.40	1.0
1-Propanol	0.10	0.9
1-Butanol	0.14	1.1
1-Hexanol	0.06	0.9
Ethylene glycol	30.00	0.4

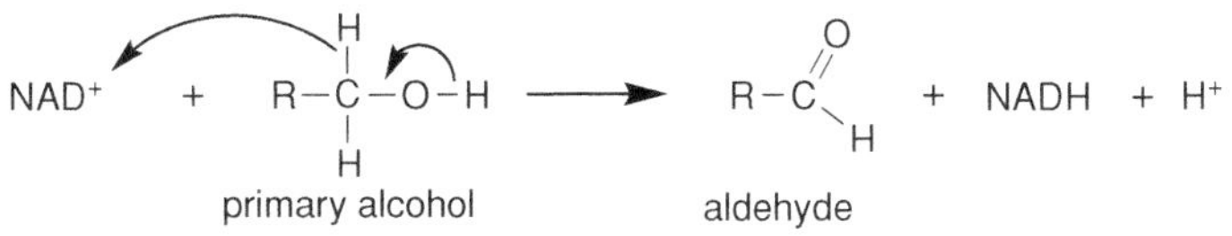

FIGURE 4.29 Mechanism of ALD-mediated oxidation of alcohols.

enriched sulindac. In contrast, the S-oxidation of the antibiotic, clindamycin, to the sulfoxide metabolite is catalyzed solely by CYP3A4 (Fig. 4.28).

Alcohol Dehydrogenase

The ALDs are a subset of the superfamily of medium-chain dehydrogenases/reductases (MDR). They are widely distributed, cytosolic, zinc-containing enzymes that utilize the pyridine nucleotide [NAD(P)$^+$] as the catalytic cofactor to reversibly catalyze the oxidation of alcohols to aldehydes in a variety of substrates. Both endobiotic and xenobiotic alcohols can serve as substrates. Examples include (72) ethanol, retinol, other aliphatic alcohols, lipid peroxidation products, and hydroxysteroids (73).

The mechanism of this oxidation is shown in Figure 4.29. The preferred cofactor for this reaction is nicotinamide adenine dinucleotide (NAD$^+$). It can be seen from this mechanism that oxidation of tertiary alcohols does not occur because there is no hydrogen on the OH-substituted carbon.

The kinetic parameters for the oxidation of a series of alcohols by ALD are shown in Table 4.1 (74). Methanol and ethylene glycol are toxic because of their oxidation products (formaldehyde and formic acid for methanol and a series of intermediates leading to oxalic acid for ethylene glycol), and the fact that their affinity for ALD is lower than that for ethanol can be used for the treatment of ingestion of these agents. Treatment of such patients with ethanol inhibits the oxidation of methanol and ethylene glycol (competitive inhibition) and shifts more of the clearance to renal clearance thus decreasing toxicity. ALD is also inhibited by 4-methylpyrazole.

Aldehyde Dehydrogenases

The aldehyde dehydrogenases are members of a superfamily of pyridine nucleotide [NAD(P)$^+$]-dependant oxidoreductases that catalyze the oxidation of aldehydes to

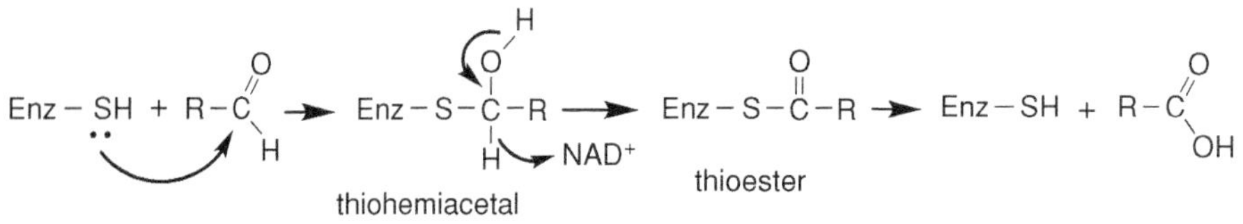

FIGURE 4.30 Mechanism of aldehyde dehydrogenase–mediated oxidation of aldehydes.

carboxylic acids. Seventeen genes have been identified in the human genome that code for aldehyde dehydrogenases attesting to the importance of these enzymes to normal physiological function. They are widely distributed and are found in cytosol, mitochondria, and microsomes (75).

As in the oxidation of alcohols, the reaction involves the loss of two hydrogen atoms rather than the addition of an oxygen atom. The mechanism of the oxidation mediated by aldehyde dehydrogenase is similar to that of ALD, but first the enzyme must form a thiohemiacetal with the substrate to facilitate the loss of hydride (76) as illustrated in the following reaction sequence (Fig. 4.30).

Most inhibitors of aldehyde dehydrogenase are inhibitors because they react with the thiol group at the active site of the enzyme. Inhibitors such as disulfiram (Fig. 4.31) have been used in the treatment of alcoholism because if someone drinks alcohol while taking the inhibitor, there is a buildup of acetaldehyde, which causes many unpleasant symptoms such as flushing and nausea (77,78). However, if someone drinks a large amount of alcohol while taking disulfiram it can lead to a life-threatening reaction.

Two broad categories of aldehyde dehydrogenases can be defined as (1) those that are highly substrate selective and critical for normal development and (2) those that have broader substrate selectivities and serve to protect the organism from potentially toxic aldehydes contained in food or those generated from xenobiotics (79). The aldehyde dehydrogenases–catalyzed oxidation of retinal to retinoic acid (Fig. 4.31), a molecule important for growth and development, falls in the first category (80). Clearly, a deficiency in this category of enzymes would be of major consequence to the organism.

The second category of aldehyde dehydrogenases are efficient catalysts of the oxidation of both aryl and alkyl aldehydes to the corresponding carboxylic acids. The most well known and common of such reactions is the oxidation of acetaldehyde, derived from alcohol, to acetic acid.

There are virtually no drugs that are aldehydes but there are dietary aldehydes, and aldehydes are common metabolites of drugs including the aldehydes generated as intermediates in monoamine oxidase (MAO)–catalyzed deamination reactions, P450-catalyzed

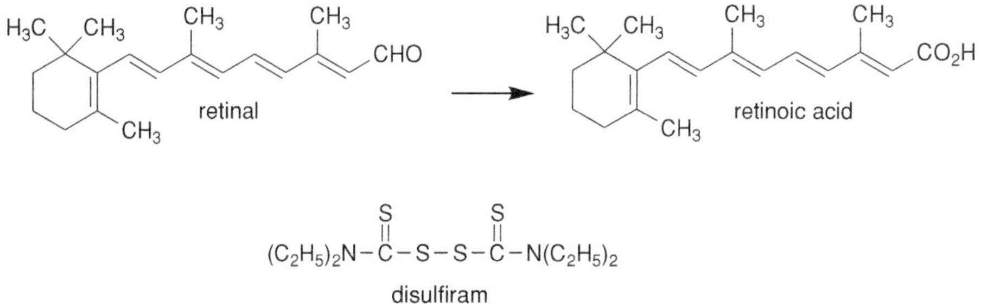

FIGURE 4.31 Conversion of retinal to retinoic acid mediated by aldehyde dehydrogenase and the structure of disulfiram, an inhibitor of aldehyde dehydrogenase.

N-dealkylation, O-dealkylation, oxidative dehalogenation, and oxidation of aryl and alkyl methyl groups.

Monoamine Oxidase

The biogenic amines are the preferred substrates of MAO. The enzyme comes in two flavors, MAO-A and MAO-B, both of which, like FMO, rely on the redox properties of FAD for their oxidative machinery. The two isoforms share a sequence homology of approximately 70% (81) and are found in the outer mitochondrial membrane, but they differ in substrate selectivity and tissue distribution. In mammalian tissues MAO-A is located primarily in the placenta, gut, and liver, while MAO-B is predominant in the brain, liver, and platelets. MAO-A is selective for serotonin and norepinephrine and is selectively inhibited by the mechanism-based inhibitor clorgyline (82). MAO-B is selective for β-phenethylamine and tryptamine, and it is selectively inhibited by the mechanism-based inhibitors, deprenyl and pargyline (82) (Fig. 4.32). Recently, both MAO-A (83) and MAO-B (84) were structurally characterized by x-ray crystallography.

The catalytic activity of MAO can be simply characterized as two half-reactions (Fig. 4.33). In the first half-reaction, the amine substrate is oxidized and the FAD cofactor is reduced. In the second half-reaction, the imine product is released and the FAD cofactor reoxidized generating peroxide. The released imine chemically hydrolyzes to the corresponding aldehyde.

The reaction is generally believed to be initiated by a single electron transfer (SET) mechanism (Fig. 4.34A). A lone-pair electron is transferred from the amine to FAD to form an aminyl radical cation and the FAD radical anion. The FAD radical anion is protonated to form the FADH semiquinone radical. The exact pathway by which the radical cation is converted to the imine is unknown but two possibilities, labeled 1 and 2, are shown in Figure 4.34A. Despite the general acceptance of the SET mechanism, it has not been definitively established and a few perplexing problems remain. For example, the existence of any intermediates expected of the SET pathway have not been confirmed by spectral data; energy calculations suggest that SET from the amine to FAD is unlikely, and deuterium isotope effects that have been determined are more consistent with a hydrogen atom transfer (HAT) mechanism (85) (Fig. 4.34B).

In the HAT mechanism, a hydrogen atom is transferred from the α-carbon of the amine to FAD to form a carbinyl radical and FADH semiquinone radical. FADH semiquinone radical abstracts the unpaired electron from the carbinyl radical forming the imine and $FADH_2$ after protonation.

A polar nucleophilic mechanism has also been advanced (86) (Fig. 4.34C). The mechanism is characterized by a nucleophilic attack of the amine on the 4α position of FAD to form the amine adduct followed by base-catalyzed elimination to the imine and $FADH_2$.

While MAO does not, in general, appear to play a major role in the metabolism of amine-containing drugs, there is at least one well-documented case of the biotransformation of a nonbiogenic amine. N-methyl-4-phenyl-1,2,3,6-tetrahydropyridine (MPTP) (Fig. 4.35), a contaminant generated in the synthesis of an illegal street drug, is oxidized by MAO-B (87) to N-methyl-4-phenyl-2,3-dihydropyridine ($MPDP^+$), the intermediate on the path to formation of N-methyl-4-phenylpyridine (MPP^+), the neurotoxin that causes a Parkinson-like syndrome (see also Chapter 8). The reaction is not restricted to MPTP alone as MAO-B will oxidize a number of MPTP analogs in which the N-methyl group has been altered and/or the 4-aryl substituent (88).

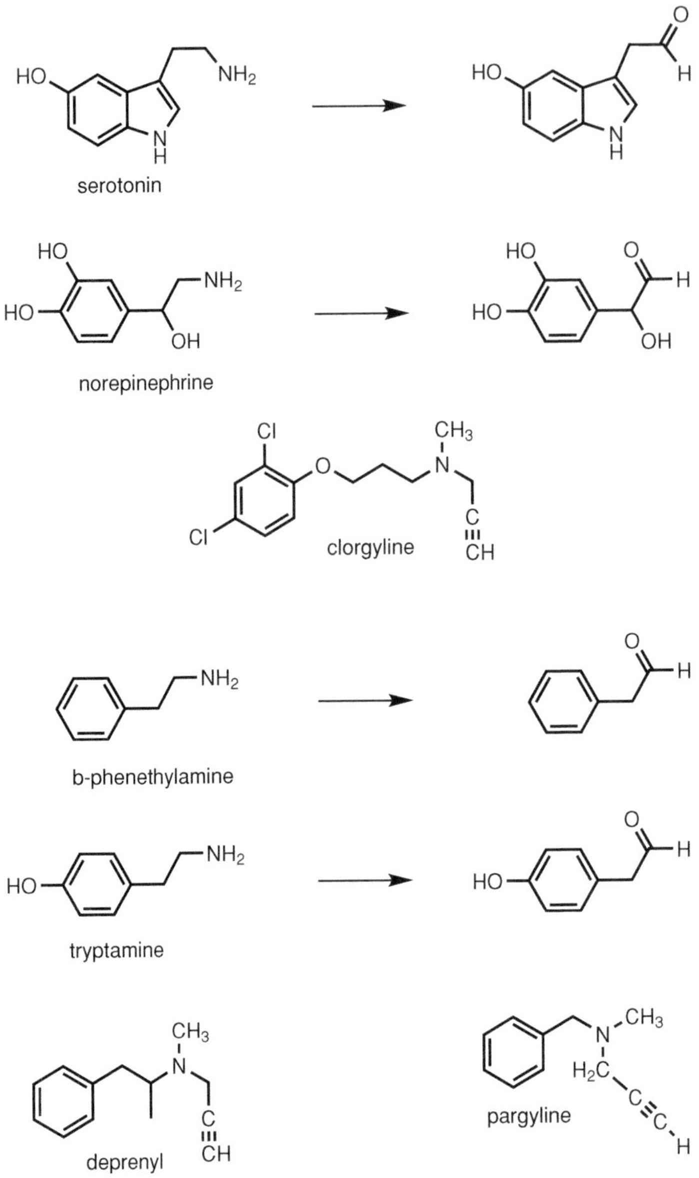

FIGURE 4.32 Structures of the MAO-A substrates (serotonin, norepinephrine), their deaminated metabolites, and its selective inhibitor, clorgyline, as well as the structures of the MAO-B substrates (β-phenethylamine, trytamine), their deaminated metabolites, and its selective inhibitors, deprenyl and pargyline.

(A) $RCH_2NH_2 + E\text{-}FAD \longrightarrow RCH{=}NH + E\text{-}FADH_2$

(B) $RCH{=}NH + E\text{-}FADH_2 \xrightarrow{O_2 \quad H_2O_2} RCH{=}NH + E\text{-}FAD$

(C) $RCH{=}NH + H_2O \longrightarrow RCH{=}O + NH_3$

FIGURE 4.33 Reactions (A) and (B) describe the MAO catalytic cycle while reaction (C) describes the subsequent aminolysis of the MAO-produced imine product to the aldehyde.

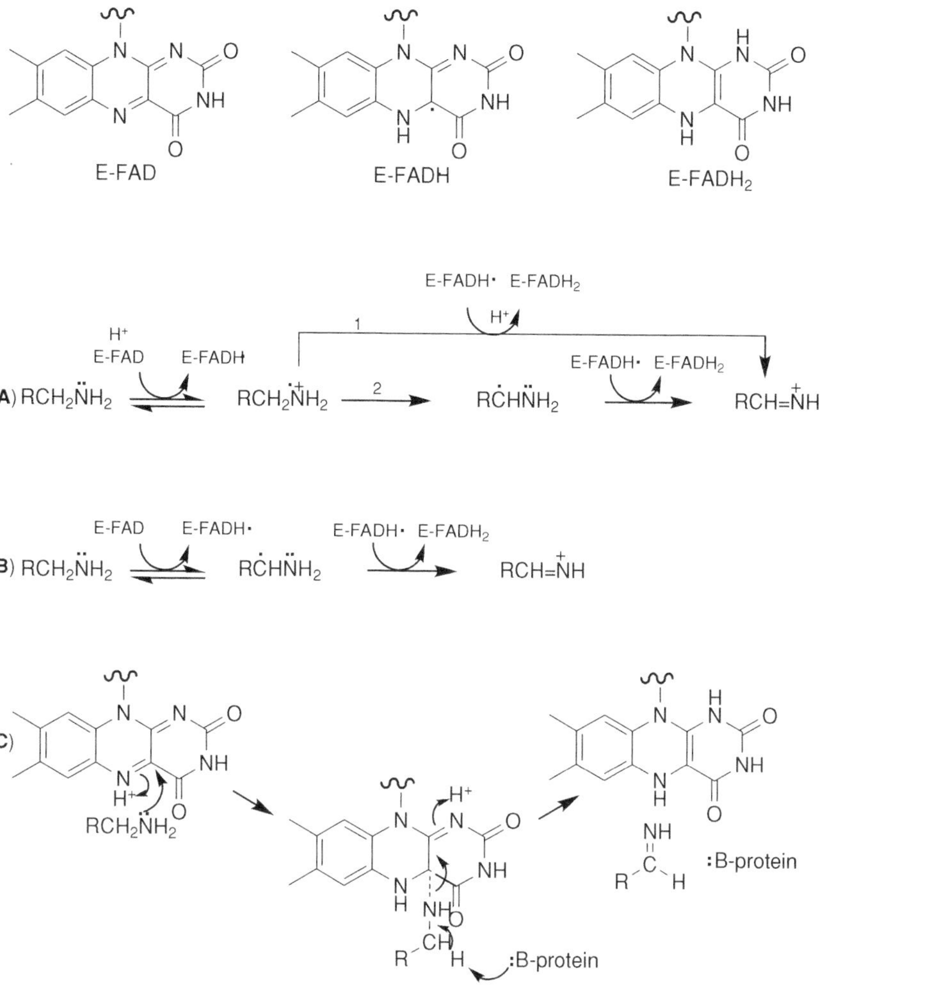

FIGURE 4.34 Postulated mechanisms for MAO-catalyzed oxidation of primary amines, (A) SET (B) HAT mechanism, and (C) polar mechanism.

The extent to which MAOA or MAOB contributes to the metabolism of other amine-containing compounds or drugs has not been determined.

Xanthine Oxidase

Xanthine oxidoreductase (XOR) is a molybdenum-containing complex homodimeric 300-kDa cytosolic enzyme. Each subunit contains a molybdopterin cofactor, two non-identical iron–sulfur centers, and FAD (89). The enzyme has an important physiologic role in the oxidative metabolism of purines, e.g., it catalyzes the sequence of reactions that convert hypoxanthine to xanthine then to uric acid (Fig. 4.36).

The oxidative reaction catalyzed by XOR is unusual relative to most oxidative enzymes, certainly P450, in that a molecule of water is the source of the oxygen atom that is transferred to hypoxanthine rather than a molecule of oxygen (90). This means that the overall reaction provides electrons rather than consuming them. The stoichiometry

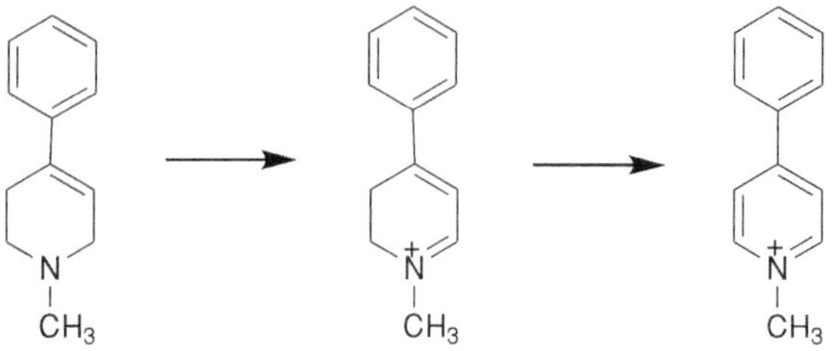

FIGURE 4.35 MAO-mediated oxidation of MPDP$^+$ to the toxic MPP$^+$.

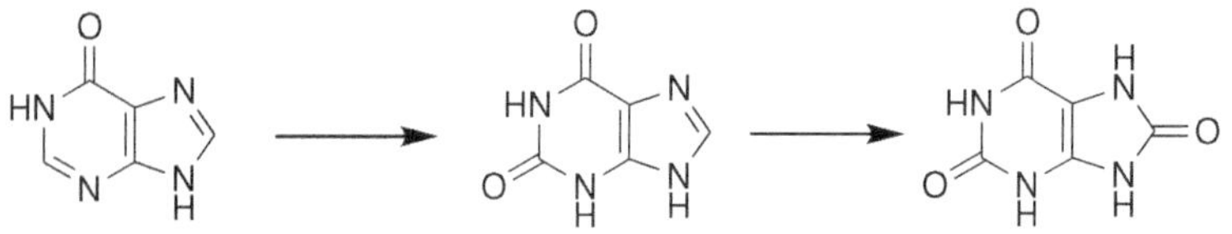

FIGURE 4.36 Xanthine oxidase–mediated oxidation of hypoxanthine to uric acid.

$$SH + H_2O \longrightarrow SOH + 2e^- + H^+$$

$$2O_2 + 2e^- \longrightarrow 2O_2^{\cdot-}$$

$$O_2 + 2e^- + 2H^+ \longrightarrow H_2O_2$$

FIGURE 4.37 Stoichiometry for XOR-catalyzed oxidation reactions.

of the general reaction is indicated in Figure 4.37, and as can be seen, the overall reaction produces electrons. The electrons gained through oxidation of a water molecule by the molybdenum cofactor to form the active oxidizing species are ultimately transferred to molecular oxygen via the FAD and iron–sulfur active site components to form either superoxide anion or hydrogen peroxide as shown.

The ligands bound to the active site molybdenum center consist of two thiolates (from cysteine), a thione, an one (i.e., double bond oxygen atom), and a hydroxy group (Fig. 4.38). The one group was thought to be the active oxygen transferred to substrate, but recent evidence suggests that it is the hydroxy group (91). An active site base first ionizes the hydroxy group which then attacks and adducts to an electron-deficient site on the substrate. In the process (Fig. 4.38), hydride is transferred from the attack site to the thione of the metal center reducing it to thiol and Mo^{VI} to Mo^{IV}. The newly formed thiol is oxidized back to thione converting Mo^{IV} to Mo^{VI} and transferring two electrons to the active-site redox components then on to molecular oxygen. The Mo^{VI}–oxygen substrate bond is hydrolyzed releasing oxidized substrate and Mo^{VI}. Mo^{VI} is now primed to restart the cycle.

XOR is a cytoplasmic enzyme and a ready source of electrons for transfer to molecular oxygen to form reactive oxygen species such as superoxide and peroxide. It is therefore thought to be involved in free radical-generated tissue injury and has been implicated in the pathogenesis of ischemia-reperfusion damage. Moreover, it has recently been implicated in the production of peroxynitrite (89), and carbonate radical anion (92), both potent biological oxidants. Its exact role in lipid peroxidation, inflammation, and infection needs

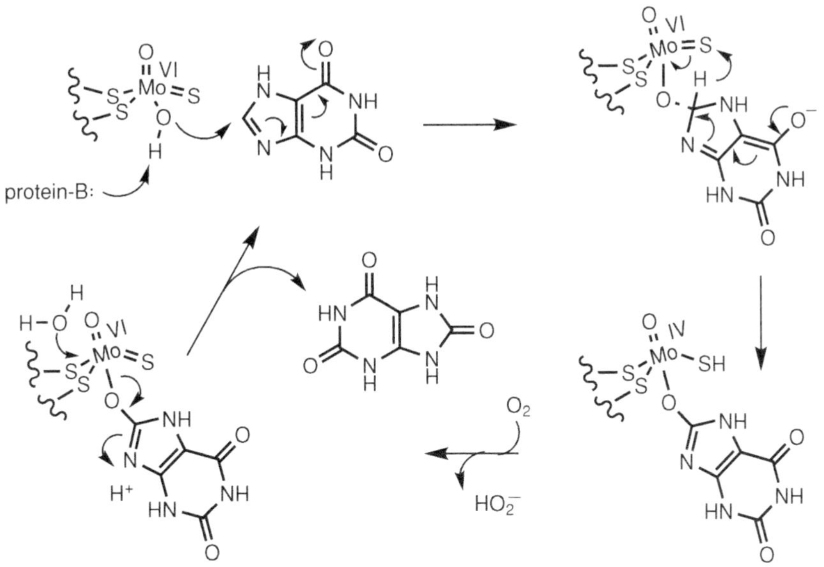

FIGURE 4.38   Mechanism for the XOR-catalyzed oxidation of purines.

to be understood as does its contribution to drug metabolism. It seems almost certain that it is likely to have a significant role in the metabolism of anticancer and antiviral agents that are structural analogs of purine or closely related heterocycles.

Aldehyde Oxidase

Aldehyde oxidase (AO) is closely related to XOR and, like XOR, is a member of the structurally related molybdo-flavoenzymes that require a molybdopterin cofactor and FAD for their catalytic activity. While AO and XOR have overlapping substrate selectivity and operate by the same chemical mechanism (93), AO has a broader substrate selectivity and as a consequence will likely play a more important role in drug metabolism. AO contributes to the oxidation of acetaldehyde resulting from alcohol ingestion. Indeed, one of the primary reactions that it catalyzes is the oxidation of aldehydes to carboxylic acids. Unfortunately, it also appears to be implicated in ethanol-induced liver injury because of the free radicals it generates in the process of oxidizing acetaldehyde and producing electrons. In oxidizing aldehydes, AO appears to operate by the same mechanism it utilizes to oxidize purines. This is illustrated for the conversion of acetaldehyde to acetic acid (Fig. 4.39).

In a recent investigation, AO, partially purified from pig liver, was found to be highly efficient ($V_{max}/K_m = 10 - 73$ mL/min/mg enzyme) in catalyzing the oxidation of a series of 11 methoxy- and hydroxy-substituted benzaldehydes (94). The fact that pig-liver AO substrate selectivity closely tracks that of human AO suggests that aldehyde AO activity might be a significant factor in the oxidation of the aromatic aldehydes generated from amines and alkyl benzenes during drug metabolism. XOR was also found to catalyze the same reactions but with much less efficiency.

AO is also effective in metabolizing a wide range of nitrogen-containing heterocycles such as purines, pyrimidines, pteridines, quinolines, and diazanaphthalenes (95). For example, phthalazine is rapidly converted to 1-phthalazinone by AO and the prodrug, 5-ethynyl-2-(1H)-pyrimidone, is oxidized to the dihydropyrimidine dehydrogenase mechanism–based inhibitor, 5-ethynyluracil, by AO (Fig. 4.40) (96).

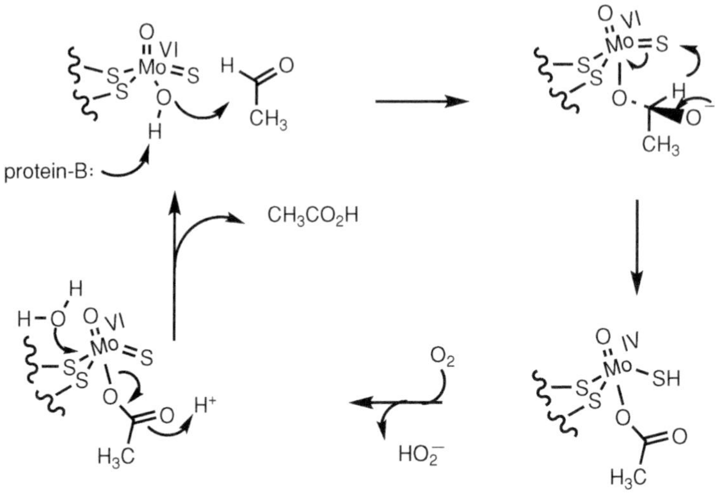

FIGURE 4.39 AO-catalyzed oxidation of acetaldehyde to acetic acid.

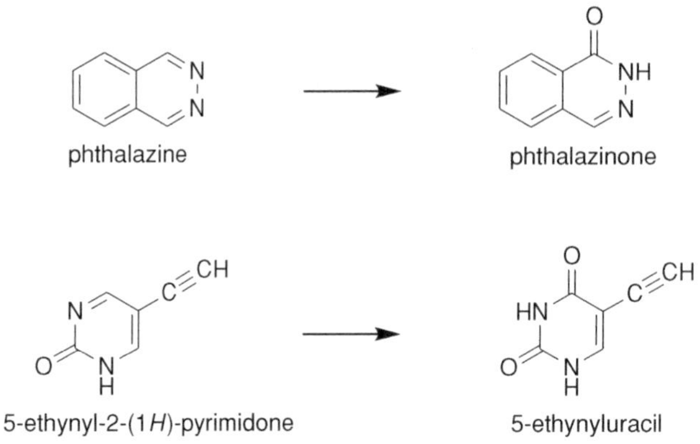

phthalazine

phthalazinone

5-ethynyl-2-(1*H*)-pyrimidone

5-ethynyluracil

FIGURE 4.40 Structures of the AO substrates, phthalazine and 5-ethynyl-2-(1*H*)-pyrimidone, and their oxidized metabolites, phthalazinone and 5-ethynyluracil, respectively.

As more information becomes available, it is becoming increasingly probable that the contributions of AO to the metabolism of a select group of drugs, those that contain aryl nitrogen–containing heterocycles, have been largely unrecognized and underestimated.

OXIDATIVE PATHWAYS

In this section, the oxidation pathways are organized by functional groups analogous to the organization of most organic chemistry textbooks thus making it easier to find what oxidative metabolic pathway would be expected at a specific site on a drug molecule.

Oxidation of sp³ Carbon–Hydrogen Bonds of Simple Alkanes

The oxidation of alkanes involves what is formally the insertion of an oxygen atom into a carbon–hydrogen bond (Fig. 4.41), although the reality of the mechanism is considerably more complex.

$$R - \overset{|}{\underset{|}{C}} - H \longrightarrow R - \overset{|}{\underset{|}{C}} - OH$$

FIGURE 4.41 General scheme for the oxidation of a C–H bond.

Cytochromes P450 are the only mammalian metabolic enzymes that can oxidize simple alkanes and the mechanism of this reaction was discussed early in this chapter. As was seen, the oxidation of carbon–hydrogen bonds can generally be rationalized as proceeding by a two-step pathway. The first step involves abstraction of a hydrogen atom by FeO^{3+} to form the $Fe^{3+} \cdot OH$ stabilized hydroxy radical and a carbon-based radical. In the second step, the hydroxy radical recombines with the carbon radical to generate hydroxylated product and regenerated enzyme. However, there appear to be exceptions, particularly in strained systems, where a mechanism that invokes the direct insertion of an oxygen into a carbon–hydrogen bond best explains experimental results. This seeming paradox has been resolved theoretically by a two-state reactivity paradigm; one state proceeds in two steps that involve formation of a discrete intermediate, while the second proceeds in a nonsynchronous manner that does not involve a discrete intermediate. The theory states that basically the energetics for the two distinct pathways are close and cross-over from one path to the other can occur. Which pathway dominates in any given case depends on the structure of the substrate and the enzyme.

Despite the apparent mechanistic complexity for P450-catalyzed hydrogen atom abstraction from sp^3 carbon, the electron deficient character of the resultant carbon-based radical or radical-like species generated by the two-state reactivity paradigm as described earlier suggests that, when potential resonance-stabilizing effects are comparable, the ease of formation of either of these species should mirror the relative ease of formation of the much more thoroughly studied electrophiles, carbocations. That is, hydroxylated product formation should follow the order tertiary > secondary > primary. Experimentally, this is indeed what is found. Even though they are the most sterically hindered, tertiary carbons tend to be preferentially hydroxylated. Also of significance is the preponderance of secondary alcohol formed at the secondary carbon site immediately adjacent to a terminal methyl group, i.e., $\omega - 1$ (the terminal position of an alkane is referred to as the ω position and therefore the next to the last position is referred to as the $\omega - 1$ position). This is often the major metabolic site for hydroxylation of hydrocarbon side chains in drug molecules. In contrast, hydroxylation at secondary carbon sites further removed from the terminal or ω position, $\omega - 2$, $\omega - 3$, etc. are much less significant, presumably because of the increased steric hindrance encountered at these sites.

While tertiary > secondary > primary is the order generally followed by the individual enzymes of the CYP1, 2, and 3 families, which are the major catalysts of human drug oxidation, the CYP4 family contains specialized P450s that are only marginally involved in drug metabolism. These enzymes are selective for ω hydroxylation, particularly in relation to the metabolism of fatty acids. For example, CYP4A11, a P450 isolated from human liver, has been identified as the major lauric acid ω-hydroxylase (97); CYP4A7, a P450 isolated from rabbit kidney, has been found to hydroxylate the prostaglandins, PGA_1 and PGA_2 (Fig. 4.42) exclusively in the ω position (98); and rat brain tissue has high fatty acid ω-hydroxylase activity leading to the formation of dicarboxylic acids (99).

It has also been shown that the CYP4 family contains a number of ω-hydroxylases whose natural substrates appear to be arachidonic acid, the prostaglandins, and/or the leukotrienes. For example, CYP4F2 and CYP4F3, isolated from human liver and human leukocytes, respectively, are leukotriene B4 ω-hydroxylases (100). CYP4F2 also catalyzes the ω-hydroxylation of arachidonic acid to form 20-hydroxy-5, 8, 11, 14-eicosatetraenoic

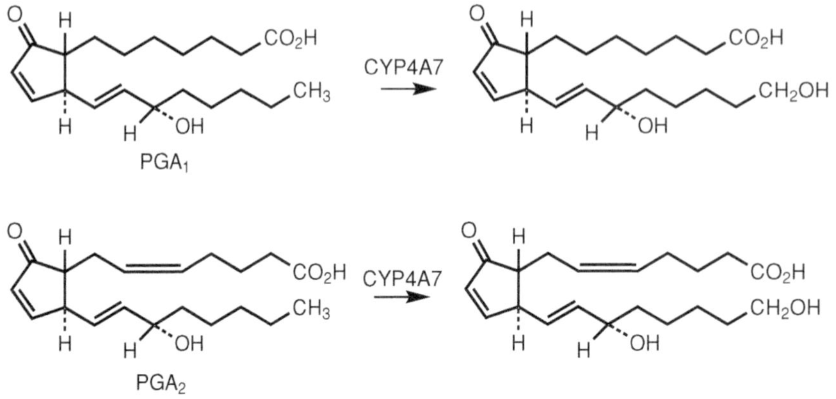

FIGURE 4.42 CYP4A7-catalyzed terminal methyl group hydroxylation of PGA$_1$ and PGA$_2$.

acid, as does human liver CYP4A11 (101), whereas CYP4A4 is a prostaglandin E$_1$ and arachidonic acid ω-hydroxylase isolated from rabbit lung (102) (Fig. 4.43).

In addition to members of the CYP4 family, specialized P450s localized in the adrenal cortex and/or the testes of the male or ovaries of the female, are critical to hormonal steroid production (Fig. 4.44). CYP11A1 (P450$_{scc}$) catalyzes the oxidative side chain cleavage of cholesterol, between carbons 20 and 22, to generate pregnenolone. Pregnenolone is then further oxidized and the 5,6 double bond isomerized to the conjugated 3,4 double bond to generate the female reproduction hormone, progesterone. CYP17 catalyzes the 17α hydroxylation of pregnenolone or progesterone, followed by cleavage of the C17–C20 to generate

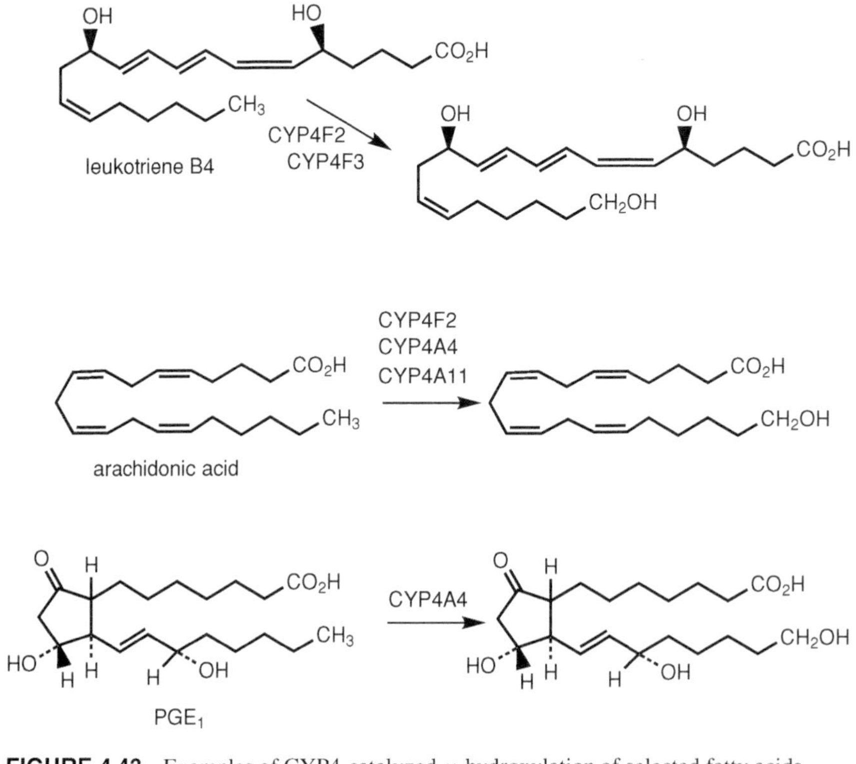

FIGURE 4.43 Examples of CYP4-catalyzed ω-hydroxylation of selected fatty acids.

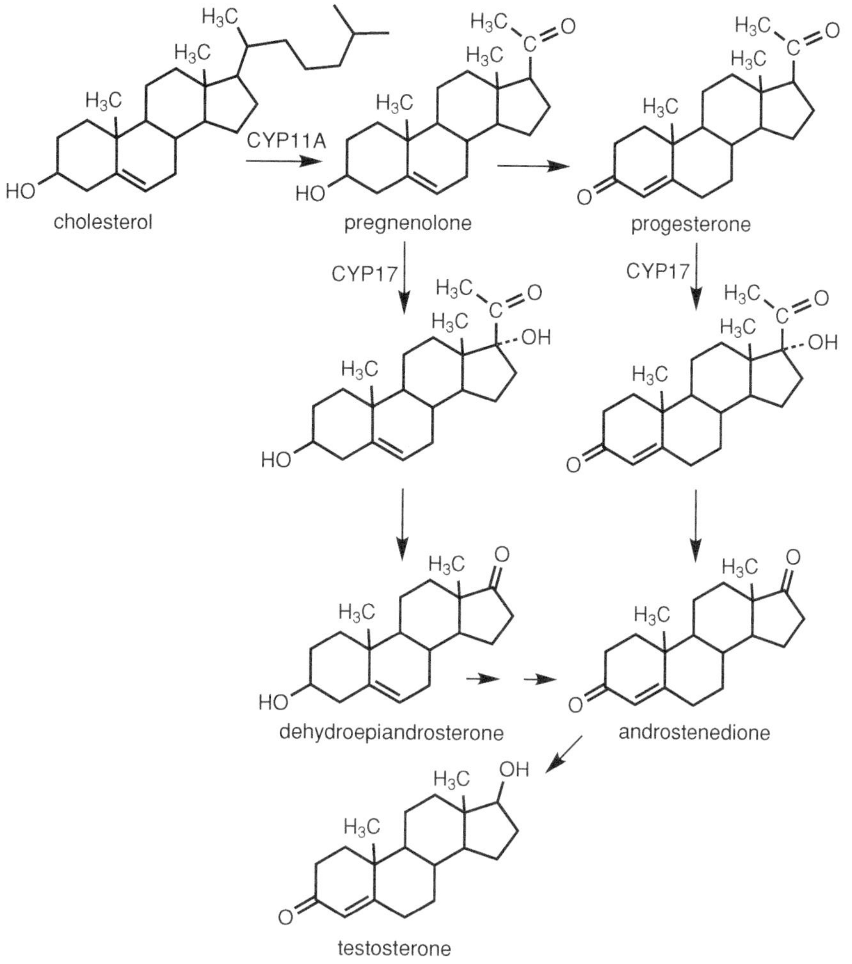

FIGURE 4.44 Cytochrome P450–dependent oxidative conversion of cholesterol to progesterone and testosterone.

dehydroepiandrosterone or androstenedione, respectively (103). Dehydroepiandrosterone can be converted to androstenedione by oxidation of the C3 hydroxy group to a ketone followed by isomerization of the 5,6 double bond in reactions analogous to those that convert pregnenolone to progesterone. Reduction of the C17 carbonyl group of androstene-dione to a 17β hydroxy group leads to the formation of the male sex hormone, testosterone. Similarly, the formation of the adrenocortical steroids, cortisol, aldosterone, the bile acids, and vitamin D all require specialized P450s at one or more steps in their biosynthetic pathways.

These results emphasize the capacity of the FeO^{3+}-activated oxygen species to selectively oxidize a carbon–hydrogen bond as unreactive as a methyl group attached to a saturated aliphatic ring or chain. They also highlight the importance of active site archi-tecture in controlling, presumably through steric interactions in the case of hydrocarbons, specifically what part of the substrate molecule is exposed to the reactive oxygen atom of FeO^{3+}. Members of CYP families 1, 2, and 3, the primary P450s involved in host defense and drug metabolism, would be expected to have relatively open and less constraining active sites in order to accommodate a greater variety of molecules, not only of differing

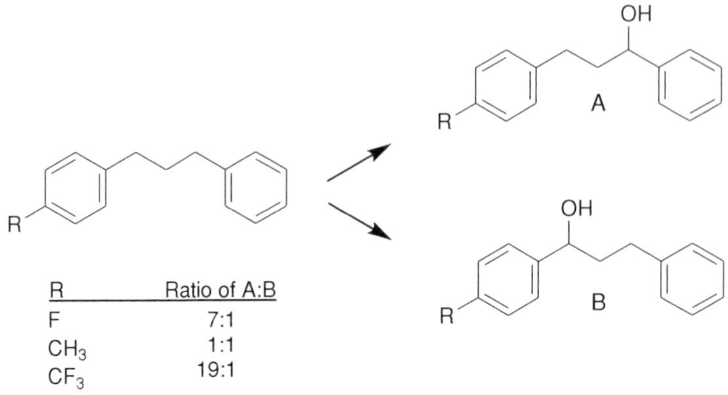

FIGURE 4.45 Cytochrome P450–catalyzed hydroxylation of *p*-mono-substituted 1,3-diphenylpropanes.

size but of differing structural type. Conversely, members of CYP families, such as CYP4, involved in both the catabolism and metabolism of endogenous bioactive molecules, such as the prostaglandins or the steroids, critical to normal physiological function, would be expected to have active sites that confine the oxidation of specific molecules to specific sites within the molecule. This indeed appears to be the case.

Oxidation of Benzylic and Allylic sp³ Carbon–Hydrogen Bonds

Benzylic and allylic carbon–hydrogen bond hydroxylation would be expected to be energetically favored processes if they proceed by the same P450-catalyzed hydrogen abstraction mechanism that operates for the simple saturated systems discussed above. Resonance stabilization of either an electrophilic transition state (developing alkyl radical) or a free-radical intermediate would lower the activation energy for both benzylic and allylic hydroxylation relative to hydroxylation of other saturated carbon–hydrogen bonds not adjacent to a π system or a heteroatom. This is generally what is found in drug metabolism studies and will be illustrated with some examples of benzylic hydroxylation.

An early study (104) in which the mechanism of benzylic hydroxylation was investigated using rat liver microsomes and several substituted 1,3-diphenylpropanes as substrates provided strong evidence that the reaction is electrophilic in character (Fig. 4.45). These particular substrates were deliberately chosen to avoid the potential pitfalls that might be associated with simply determining the relative rates of benzyl alcohol formation from a series of para-substituted toluenes. The authors point out that the overall enzymatic velocity might not just reflect differences in the rate of the hydroxylation step; it is also likely that different substituents will modulate the affinity of each substrate for the active site because of differing steric or electronic requirements and thus also contribute to observed changes in rate. Both potential problems were avoided by the symmetry of the substrates chosen. Each substrate molecule was designed such that a direct intramolecular competition was introduced between two benzylic sites. The enzyme is presented with an equal choice of hydroxylating an unsubstituted benzylic site or an equivalent para-substituted benzylic site. It is clear from the ratio of the two products, A:B, that reaction preferentially occurs at the benzylic site where electron deficiency can best be stabilized.

In a more recent study (105), the intramolecular deuterium isotope effect and the relative rates of benzylic hydroxylation for six para-substituted (OCH₃, CH₂D, H, Cl, Br,

TABLE 4.2 Intramolecular deuterium isotope effects for benzylic hydroxylation of substituted toluenes catalyzed by CYP1A2, CYP1B1, CYP2C9, CYP2E1, and CYP101

Substituent R	Isotope effect				
	1A2	2B1	2C9	2E1	CYP101
CH_3O	4.64	3.69	4.3	4.24	4.44
CH_2D^a	5.59	6.23	5.9	5.45	6.0
H	6.1	7	n.d.[b]	6.1	n.d.
Cl	7.06	8.1	6.2	6.75	6.5
Br	6.83	8.02	6.9	6.75	8.3
CN	10.1	11.9	11.1	10.1	11.6

[a] p-xylene-α-$^2H_1$-α'-$^2H_1$ was used as substrate.
[b] not determined.

and CN) selectively deuterated toluenes with five different recombinant P450 preparations (CYPs 2E1, 2B1, 1A2, 2C9, and 101) were determined. Unfortunately the problems that had been anticipated and avoided in the earlier study were realized in this study. A good correlation between the electron-donating power of the substituents and the rate of benzylic hydroxylation was not obtained. However, the near equivalence of the isotope effect profile of each substrate over the five different P450s clearly establishes that the mechanism for benzylic hydroxylation is independent of the P450 isoform catalyzing its formation, including the bacterial enzyme, CYP101 (Table 4.2).

A stereochemical study (106) of the hydroxylation of the prochiral benzylic carbon of phenylethane with a single P450 from rabbit liver, CYP2B4 (P450$_{LM2}$), using enantiomerically pure (R)- and (S)-phenylethane-1-d as substrates also revealed some fundamental and general properties of the P450s. Findings of particular note are (1) hydroxylation

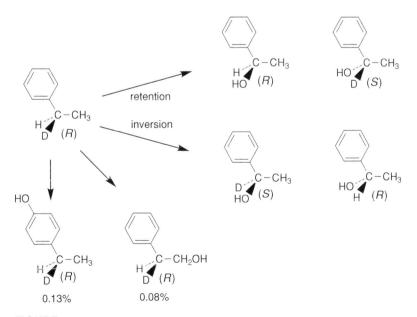

FIGURE 4.46 Stereochemistry of the cytochrome P450–catalyzed benzylic oxidation of (R)-ethylbenzene-1-d.

occurs almost exclusively, greater than 99%, at the benzylic position to give the isomeric α-methylbenzyl alcohols (Fig. 4.46), a result that strongly reinforces the notion that a benzylic site is a favored site for P450-catalyzed oxidative attack, and (2) the contribution of the minor metabolites, 2-phenylethanol and 4-ethylphenol (a trace of 2-ethylphenol is also seen in this experiment) to overall metabolism more than triples when phenylethane-1-d_2 is used as substrate (Fig. 4.46). Benzylic hydroxylation being isotopically driven to switch to secondary sites of metabolism not only indicates the operation of a significant isotope effect but the tripling in formation of minor metabolites, particularly 4-ethylphenol, also indicates that the substrate has considerable freedom of motion and the potential for forming multiple catalytically productive binding orientations within the active site (107). It is not uncommon for a single P450 isozyme, particularly those in the first three CYP families, to catalyze the formation of multiple regioisomeric products from the same parent substrate (108). In many cases, P450 substrates are hydroxylated primarily at the energetically most favored position (109).

Stereochemical analysis of the benzylic alcohols formed from both (R)- and (S)-phenylethane-1-d after incubation with CYP2B4 established a strong cross-over component

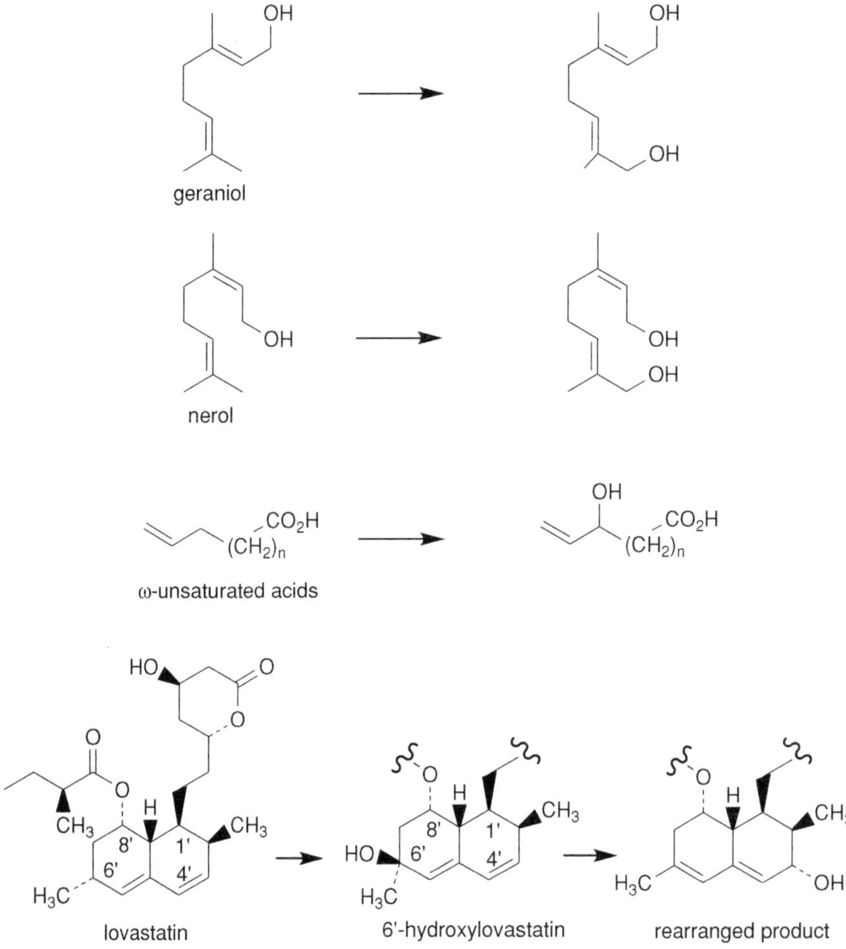

FIGURE 4.47 Examples (geraniol, nerol, ω-unsaturated acids, and lovastatin) of cytochrome P450–catalyzed allylic hydroxylation.

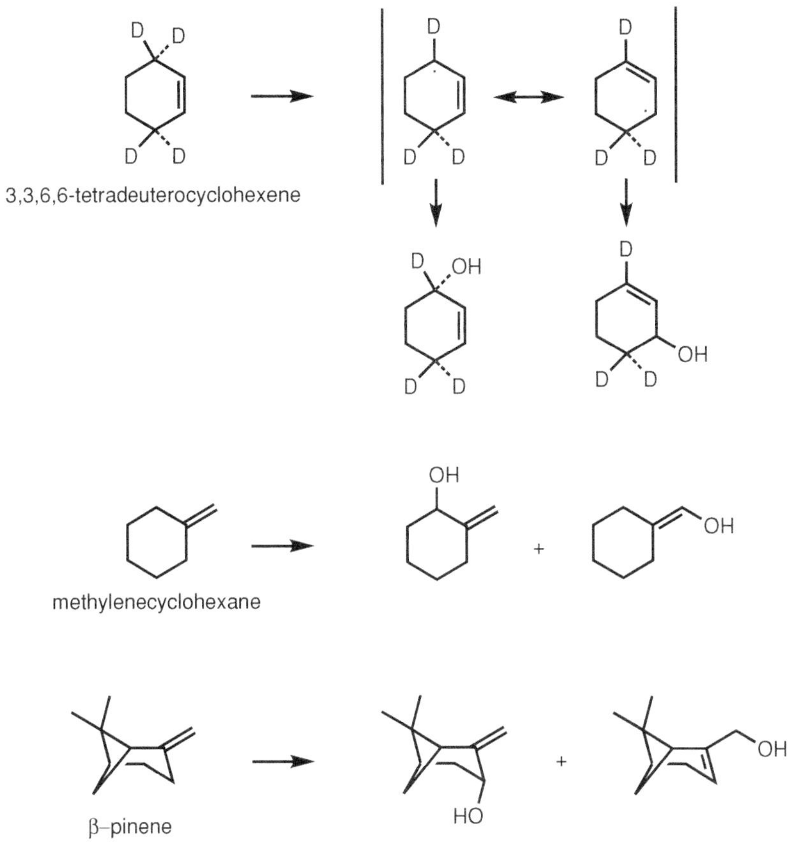

FIGURE 4.48 Substrates (3,3,6,6-tetradeuterocyclohexene, methylenecyclohexane, and β-pinene) used to determine the mechanism of cytochrome P450–catalyzed allylic hydroxylation.

in both reactions, i.e., complete retention of configuration was not observed for either substrate in the hydroxylation reaction. Lack of retention establishes the loss of stereochemical integrity in the process and necessitates the involvement of a tri-coordinate intermediate like a radical—a result consistent with the radical rebound mechanism (4) and the two-state reactivity model (8).

Allylic oxidation of the terpenes, geraniol and nerol, is particularly informative as to the nature of this type of reaction (110). Hydroxylation occurs almost exclusively at the C10 (E)-methyl group of both compounds (Fig. 4.47). Thus, the adjacency of a double bond converts a highly stable and unreactive methyl group to a major site of oxidative attack. Similarly, introduction of unsaturation (double or triple bond) converts the $\omega - 2$ position in a series of ω-unsaturated fatty acids into the major site of metabolic attack (111). A major metabolite of the cholesterol-lowering drug, lovastatin, is the 6′-hydroxy metabolite, most often found as a conjugate or the rearranged lactone ring-opened 3′-hydroxy-iso-$\Delta-4′,5′$-hydroxy acid product formed upon mild acid workup (112) (Fig. 4.47).

To explore the mechanism of allylic hydroxylation, three probe substrates, 3,3,6,6-tetradeuterocyclohexene, methylene cyclohexane, and β-pinene, were studied (113). Each substrate yielded a mixture of two allylic alcohols formed as a consequence of either retention or rearrangement of the double bond. The observation of a significant deuterium isotope effect (4–5) in the oxidation of 3,3,6,6-tetradeuterocyclohexene together with the formation of a mixture of un-rearranged and rearranged allylic alcohols from all three substrates is most consistent with a hydrogen abstraction–oxygen rebound mechanism (Fig. 4.48).

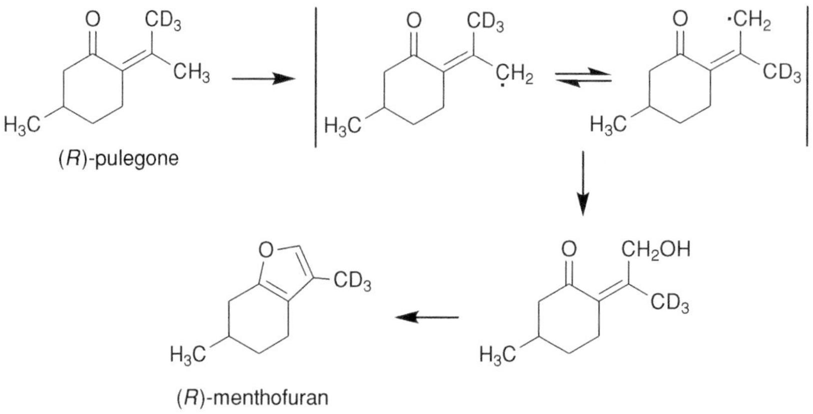

FIGURE 4.49 Mechanism for the oxidative rearrangement of (*R*)-pulegone to (*R*)-menthofuran.

A consequence of generating a radical intermediate in the P450-mediated oxidation of an allylic carbon is the possible direct production of a rearranged product as indicated above and as seen in the metabolism of (*R*)-pulegone, the major constituent of pennyroyal oil —a volatile plant oil that has been used as an abortifacient and causes major toxicity at high doses. Menthofuran, previously identified as a metabolite of pulegone, appeared to arise from initial P450-catalyzed oxidation of one of the allylic methyl groups. To probe the mechanism of the reaction, the (*E*)-methyl-d_3 analog of pulegone was synthesized and incubated with microsomal P450 to form menthofuran (114). The isolated menthofuran contained a furano-trideuteromethyl group indicating that an intermediate must have been formed during the course of the reaction to allow interchange of the positions of the two allylic methyl groups prior to hydroxylation, ring closure, and aromatization (Fig. 4.49).

Oxidation α to a Heteroatom (N, O, S, Halogen)

Given that cytochrome P450 can catalyze the oxidation of the carbon–hydrogen bonds of simple hydrocarbons, it is not surprising that they can also oxidize carbon–hydrogen bonds adjacent to heteroatoms such as nitrogen, oxygen, sulfur, or halogen. They can do this even more effectively if the heteroatom can lower the energy of activation of reaction for the formation of the incipient carbon-based radical relative to oxidation of a simple hydrocarbon.

N-Dealkylation/Deamination
The primary difference between oxidation of a carbon–hydrogen bond of a simple hydrocarbon and one adjacent to a heteroatom is that the former reaction leads to the formation of an alcohol while the latter generally leads to loss of the alkyl group. In fact, N-dealkylation is one of the most frequently encountered metabolic reactions in drug metabolism studies and is often the pathway responsible for the production of the major metabolite obtained from an *N*-alkyl-containing drug. Its prominence is not simply derived from the commonality of an alkyl-substituted amino group as an important part of the structural motif of many drugs, it is also that N-dealkylation is energetically favored (115) relative to most other metabolic pathways. A typical example is the N-deethylation of lidocaine (116) to form the secondary amine (Fig. 4.50).

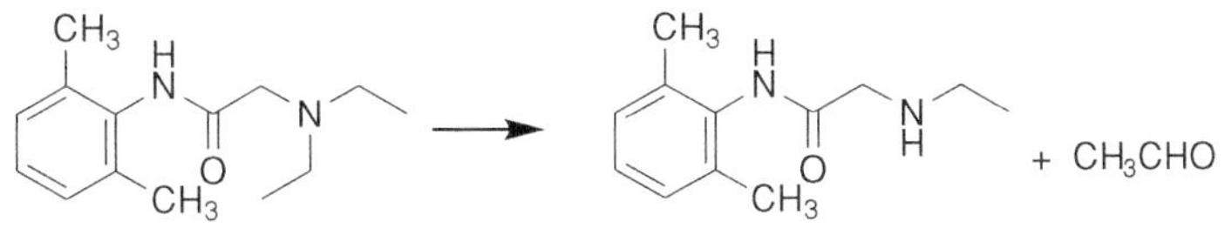

FIGURE 4.50 Cytochrome P450–catalyzed oxidative N-dealkylation of lidocaine.

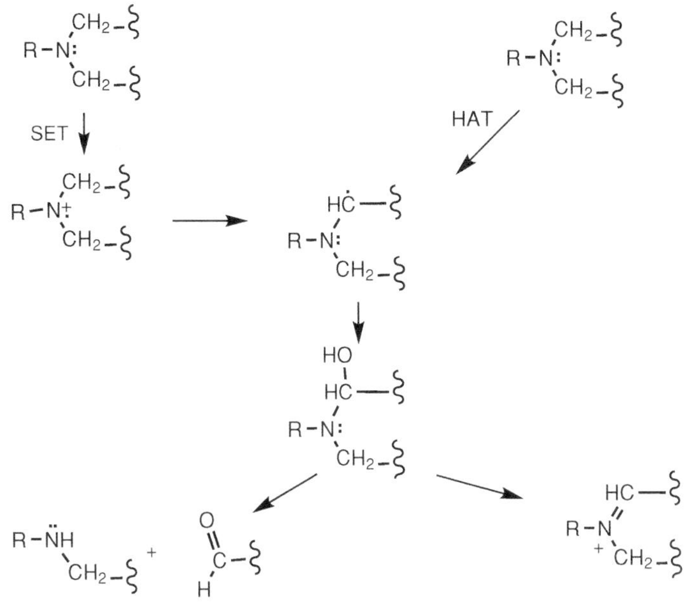

FIGURE 4.51 Mechanism for cytochrome P450–catalyzed N-dealkylation via an initial single electron pathway, SET, or via the hydrogen atom abstraction pathway, HAT.

Delineation of the exact mechanism of oxidative N-dealkylation has been problematic. Presently there are two competing mechanisms, SET (single electron transfer)-or electron-proton-electron mechanism—and the HAT (hydrogen atom transfer) mechanism (117). Both mechanisms postulate the intermediacy of a carbinolamine but differ in the mechanistic events leading to its formation (Fig. 4.51). The SET pathway is initiated by SET from the nitrogen lone pair of electrons to FeO^{3+}, which is followed by the transfer of a proton from the α-carbon to the one electron-reduced oxene to form a heme-stabilized hydroxy radical, $Fe^{+3} \cdot OH$, and the α-carbon radical. Oxygen rebound then forms the carbinolamine. The HAT pathway postulates formation of the carbinolamine by direct transfer of a hydrogen atom from the α-carbon atom to FeO^{3+} to form $Fe^{3+} \cdot OH$ followed by oxygen rebound.

Carbinolamines are chemically unstable and, in the case of tertiary amines, dissociate to generate the secondary amine and aldehydes as products or eliminate water to generate the iminium ion. The iminium ion, if formed, can reversibly add water to reform the carbinolamine or add other nucleophiles if present. If the nucleophile happens to be within the same molecule and five or six atoms removed from the electrophilic carbon of the iminium ion, cyclization can occur and form a stable 5- or 6-membered ring system. For example, the 4-imidazolidinone is a major metabolite of lidocaine, which is formed in vivo or can be formed upon isolation of the N-deethyl metabolite of lidocaine if a trace of acetaldehyde happens to be present in the solvent used for extraction (116,118) (Fig. 4.52).

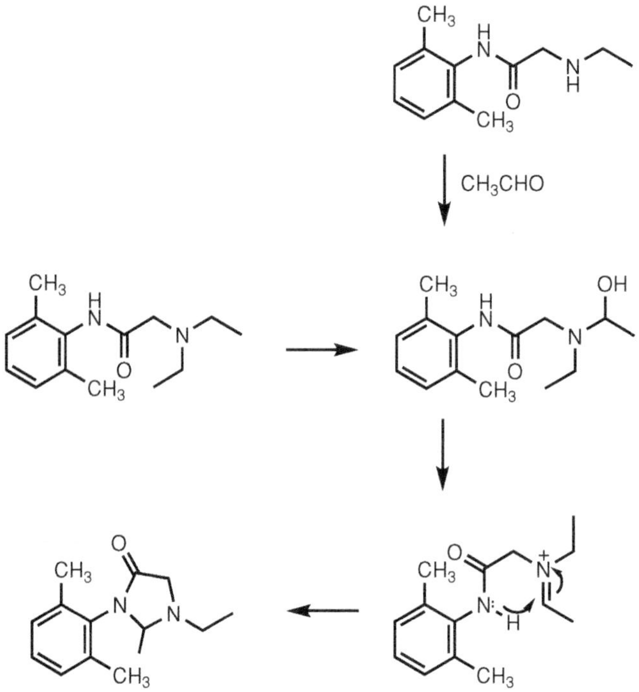

FIGURE 4.52 Mechanism for formation of the 4-imidazolidinone from the carbinolamine metabolic intermediate of lidocaine.

N-Dealkylation reactions are not restricted to tertiary amines. Secondary amines as well as primary amines can also be dealkylated although both types are less favored than tertiary amines. In the case of primary amines, the lone pair of electrons of the amino group can interact and complex with the Fe^{3+} of heme. Thus primary amines tend to be inhibitors of P450 activation and for that reason are generally poor substrates. Secondary amines have metabolic properties intermediary between those of tertiary amines and primary amines. They are less-effective inhibitors because of increased steric hindrance to complex formation but are also better substrates because they are less-effective inhibitors and thereby increase the effective concentration of enzyme.

The alkyl substituents of the amino group need not be primary to be susceptible to oxidative removal. They can be secondary like the isopropyl group of propranolol (Fig. 4.53) in which case the carbinolamine dissociates into a ketone (acetone) and a primary amine.

N-dealkylation and deamination reactions are the same reaction. For example, in the metabolism of propranolol, if oxidation occurs to the left of the nitrogen the reaction is called deamination because the nitrogen is lost from the larger part of the molecule, whereas if the oxidation occurs to the right of the nitrogen it is called a N-dealkylation because the alkyl group is removed from the nitrogen; however, the basic mechanisms of these two reactions are identical. If the two groups attached to the nitrogen were of equal size, it would be completely arbitrary whether it would be called N-dealkylation or deamination.

The disassociation of a carbinolamine is reversible, but in a biological system, when a carbinolamine dissociates to an amine and an aldehyde or ketone, these products diffuse away from each other and the dissociation is essentially irreversible. However, if the amine

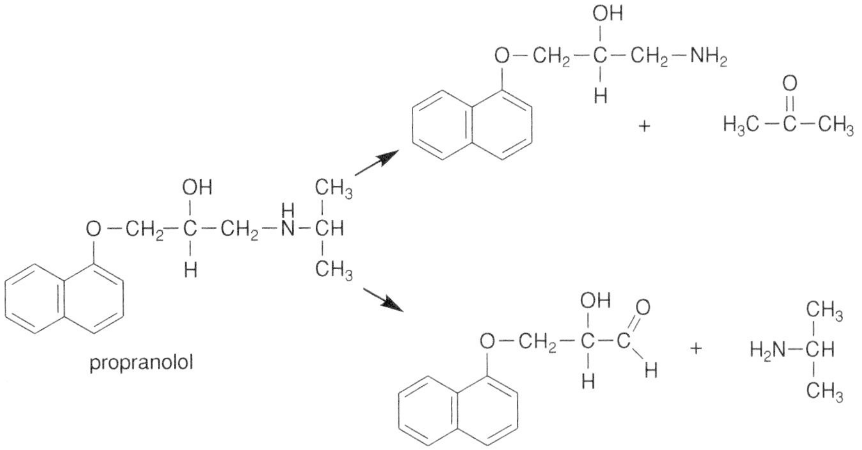

FIGURE 4.53　N-dealkylation/deamination of propranolol.

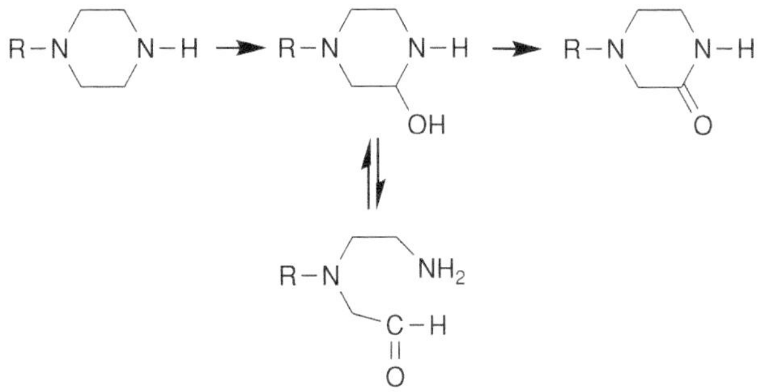

FIGURE 4.54　Further oxidation of a carbinolamine can occur if it is part of a ring system.

is part of a ring, the dissociation products are kept from diffusing away from each other and further oxidation of the carbinolamine to a lactam can occur as shown in Figure 4.54.

　　Given the mechanism of N-dealkylation/deamination, deamination of an aromatic amine or one involving a tertiary carbon should be impossible because there is no carbon–hydrogen bond that can form a carbinolamine. However, there are a few examples where such metabolic pathways are in fact observed because of special structural characteristics that allow the process. The metabolism of vesnarinone is a case in point. When mediated by myeloperoxidase, the mechanism involves N-chlorination followed by loss of HCl to form an iminium ion (see Fig. 8.15 in Chapter 8). When mediated by P450, it could also involve direct dehydrogenation to form the iminium ion (Fig. 4.55). Alternatively, α-carbon hydroxylation yields a carbinolamine, which is in equilibrium with an iminium ion in the piperazine ring (Fig. 4.55). The presence of a secondary p-amino group allows resonance-driven rearrangement of this iminium ion to the iminium ion that involves the aromatic ring and labeled iminium ion in Figure 4.55 and the key intermediate in either pathway. Addition of water to form an aryl carbinolamine followed by normal N-dealkylation leads to cleavage of the nitrogen–aryl bond (Fig. 4.55) (53). (The type of oxidation that in this case directly leads to the iminium ion will be discussed in "Oxidative Dehydrogenation" section later in this chapter.)

　　An example of N-dealkylation of an amine adjacent to a tertiary carbon can be found in the metabolism of synthetic opiod, alfentanil. The CYP3A4-catalyzed oxidation of the opiod alfentanil follows two major pathways (119): N-dealkylation to form noralfentanil and cleavage of the spiro center to generate N-phenylproprionamide. Moreover,

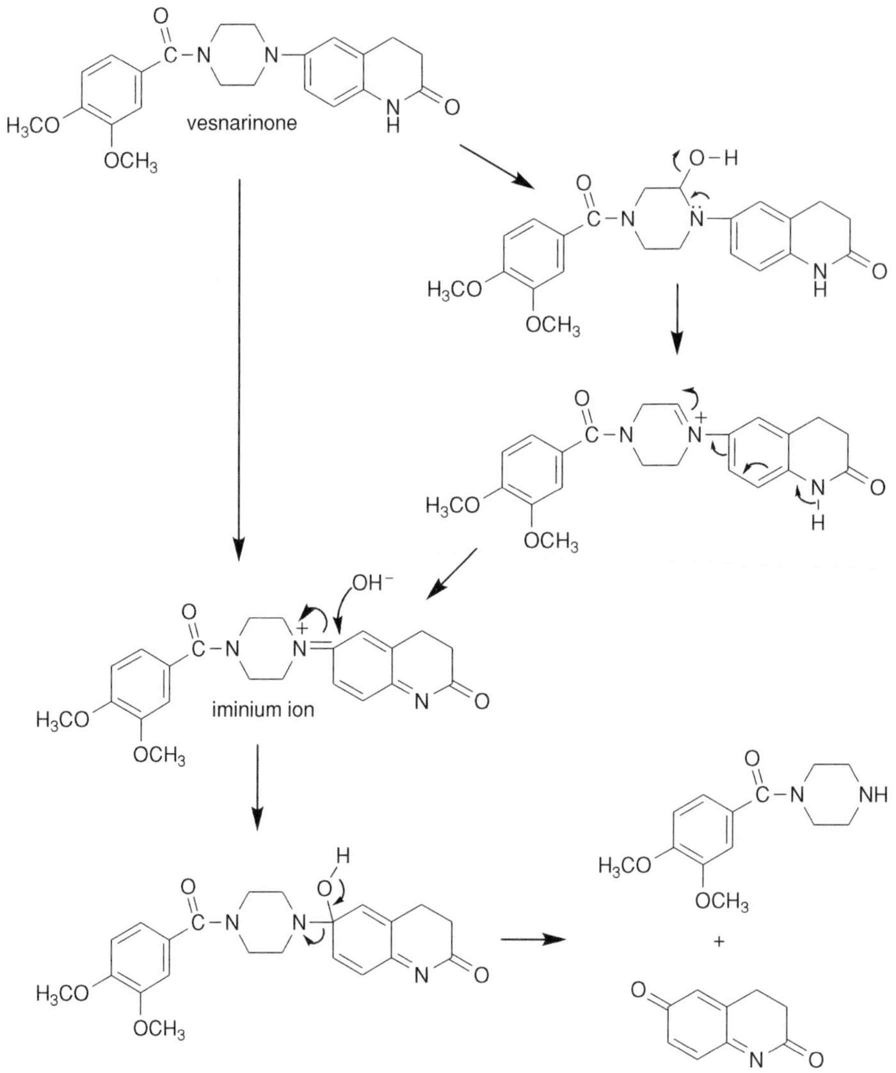

FIGURE 4.55 Deamination of an aromatic amine.

N-phenylproprionamide is found to come directly from alfentanil and not from noralfen-tanil. The mechanism of how the carbon–nitrogen of the spiro center is cleaved has not appeared in the literature. A possible mechanism would entail initial hydroxylation of a ring carbon adjacent to the piperdine nitrogen, followed by elimination of hydroxide to form the imine then rearrangement to the enamine, and finally elimination of the amide as indicated in Figure 4.56.

Oxidative attack on a carbon–hydrogen bond of an alkyl group α to a nitrogen atom is not restricted to saturated aliphatic amines. In fact X in an X–N–CH– structural subunit can be virtually any common atomic grouping that can be found in stable organic molecules. For example, α-carbon hydrogens of N-alkyl-substituted aromatic cyclic amines (119), aryl amines (120), amides (121), amidines (122), N-nitrosodialkylamines (123), etc. are all subject to oxidative attack, carbinolamine formation, and in most cases release of an aldehyde or ketone depending on the substitution pattern (1° or 2°)

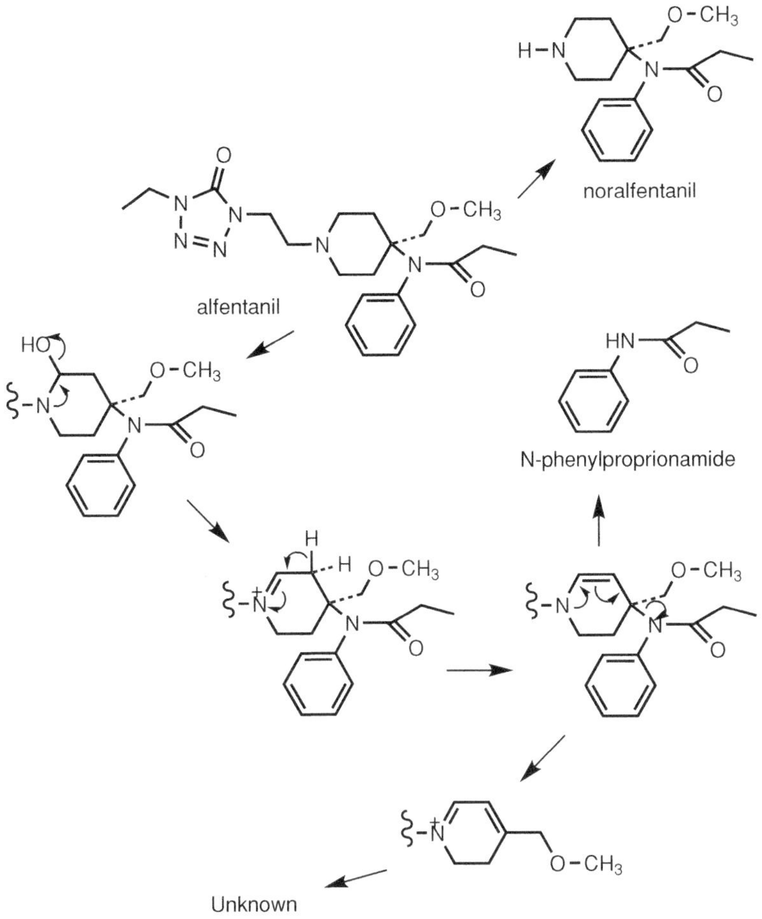

FIGURE 4.56   Possible mechanism for the direct loss of N-phenylproprionamide from alfentanil.

(Fig. 4.57). In some cases, particularly N-alkyl aromatic cyclic amines, the carbinolamines that are formed are stable enough to be isolated.

O-Dealkylation

The O-dealkylation of ethers, while not as frequently encountered as N-dealkylation in drug metabolism studies, is still a common metabolic pathway. Mechanistically it is less controversial than N-dealkylation in that it is generally believed to proceed by the HAT pathway, i.e., α-hydrogen atom abstraction followed by hydroxyl radical rebound, and not a SET pathway (Fig. 4.58). The product of the reaction is unstable, being a hemiacetal or hemiketal depending on the number of hydrogens on the α-carbon, which dissociates to generate an alcohol and an aldehyde or ketone.

Energetically, O-dealkylation is less favored than N-dealkylation (108). This is not surprising as the greater electronegativity of oxygen relative to nitrogen would make abstraction of an α-hydrogen atom more difficult. Examples of drugs in which O-dealkylation plays a significant role are propranolol (Fig. 4.53), phenacetin, dextromethorphan, codeine, and metoprolol (Fig. 4.59).

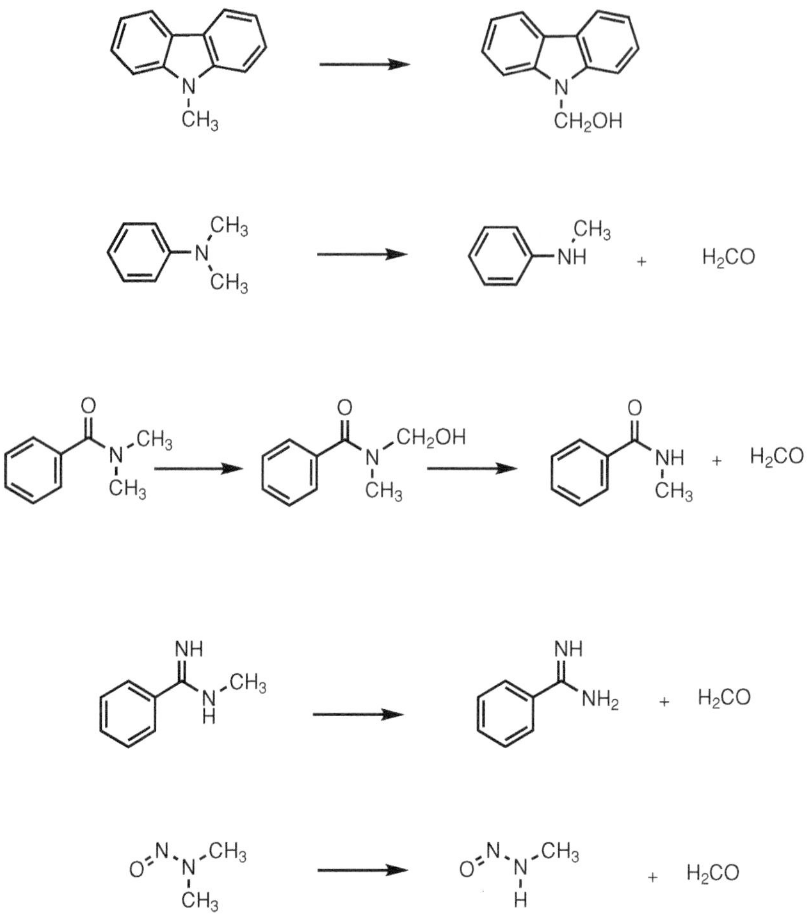

FIGURE 4.57 Examples (*N*-methylcarbazole, *N*,*N*-dimethylaniline, *N*,*N*-dimethylbenzamide, *N*-methylbenzamidine, and *N*,*N*-dimethylnitrosamine) of cytochrome P450–catalyzed oxidative N-dealkylation.

P450 can also catalyze hydroxylation of a carbon–hydrogen bond α to an oxygen atom in an alcohol. But, in contrast to the ethers, the primary oxidants of alcohols appear not to be the P450s but other enzymes like the dehydrogenases as will be discussed later.

S-Dealkylation
The basic scheme for the S-dealkylation is the same as for other dealkylations (Fig. 4.60). S-Dealkylation unlike either N- or O-dealkylation is relatively uncommon, generally not a major metabolic pathway, and in some cases might not even contribute to the overall metabolic profile of a sulfide-containing drug. This is probably due to two factors: (1) sulfide-containing drugs represent a small percentage of available drugs and (2) the sulfur atom itself is more susceptible to oxidation than is the adjacent α-carbon–hydrogen bond. Nevertheless, S-dealkylation does occur; the expected intermediate being a thiohemiacetal as shown in Figure 4.60. However, whether S-dealkylation is driven by cytochrome P450 or some other enzyme system is not clear.

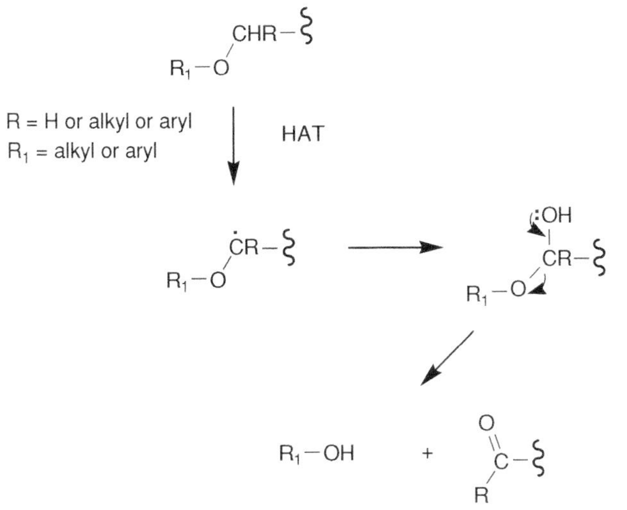

FIGURE 4.58   Mechanism for O-dealkylation via the HAT pathway.

Halogen Dealkylation

Halogen dealkylation mimics O-dealkylation both in terms of mechanism and the commonality of the process. Virtually any drug that contains a carbon–hydrogen bond adjacent to a halogen atom will be subject to P450-catalyzed oxidative dehalogenation (Fig. 4.61).

While aliphatic halogen is not a common structural component of most drugs, it is a major structural component of most inhalation anesthetics. As might be expected because halogens stabilize the carbon radical, the more halogens that are present on the carbon, the faster the oxidation is likely to be and also the less electronegative the halogen, the faster it is likely to be. Even more importantly, the presence of two halogens on a carbon leads to the metabolic production of an acid halide, a highly reactive and toxic species.

A classic example is the activation of halothane to trifluoroacetyl chloride predominantly by CYP2E1 as shown in Figure 4.62, which is responsible for its hepatotoxicity (124). Trifluoroacetyl chloride can covalently bind to protein generating liver-protein neoantigens (125). In susceptible individuals, these neoantigens stimulate production of anti trifluoroacetyl–protein antibodies. It is not clear that these antibodies are pathogenic; it may be T cells that actually destroy the liver, but the antibodies are evidence of an immune response. In some cases, this results in fatal halothane hepatitis in the individual upon reexposure to the anesthetic. The structurally related inhalation anesthetics, enflurane and isoflurane, are also subject to α-carbon hydroxylation followed by acyl halide formation and an ensuing hepatic dysfunction similar to that caused by halothane, although oxidative dehalogenation of isoflurane is less than that of halothane and the incidence of hepatotoxicity is also lower (126,127).

Aliphatic halogen is present in a number of common solvents and industrial chemicals. For example, the fuel additive and suspected human carcinogen, 1,2-dibromoethane, is oxidatively transformed to bromoacetaldehyde by CYP2E (Fig. 4.63). The aldehyde functional group of bromoacetaldehyde is chemically reactive and susceptible to addition reactions with in vivo nucleophiles, while the bromo group is also prone to displacement by a nucleophile. Thus, bromoacetaldehyde might be an efficient cross-linking agent.

A new class of compounds that has encountered problems similar to those of the inhalation anesthetics is the hydrochlorofluorohydrocarbons (HCFCs). These compounds

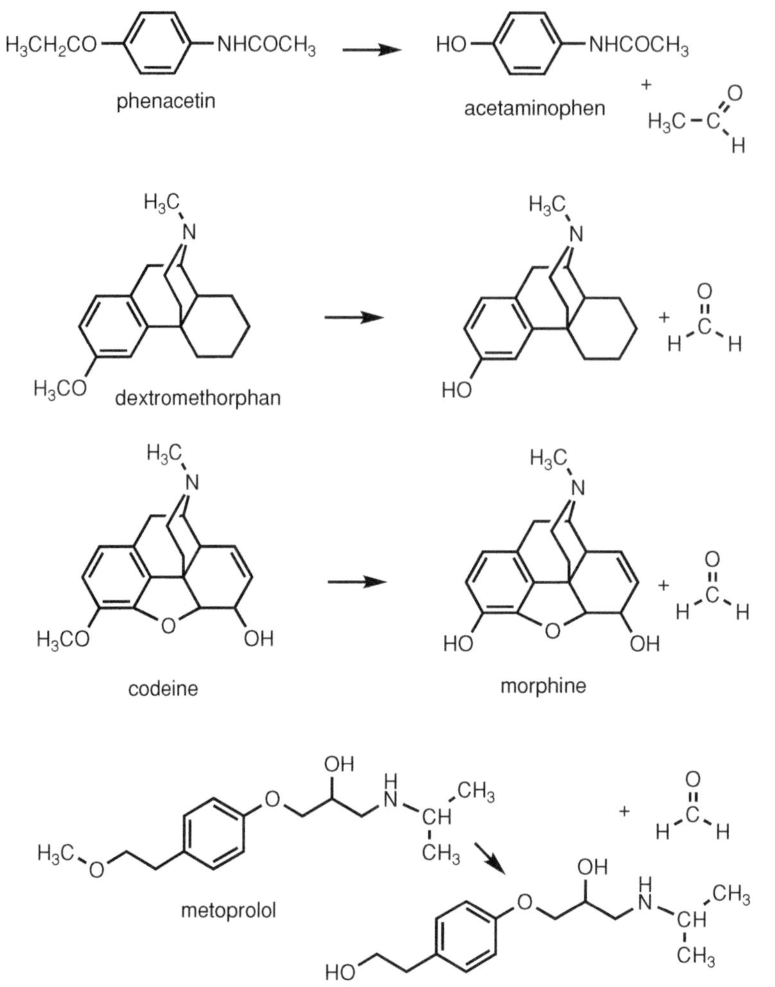

FIGURE 4.59 Examples (phenacetin, codeine, dextromethorphan, and metoprolol) of cytochrome P450–catalyzed O-dealkylation.

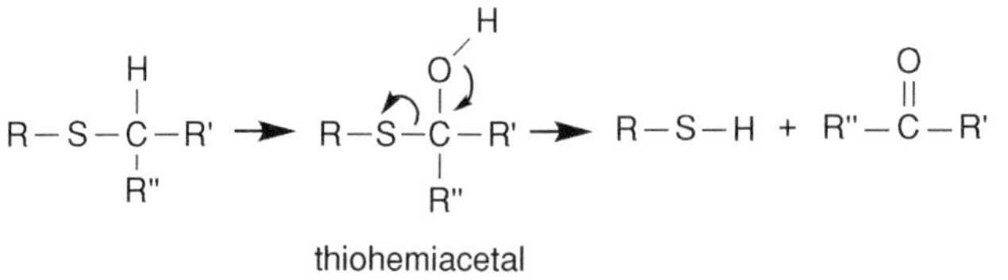

FIGURE 4.60 General scheme for S-dealkylation.

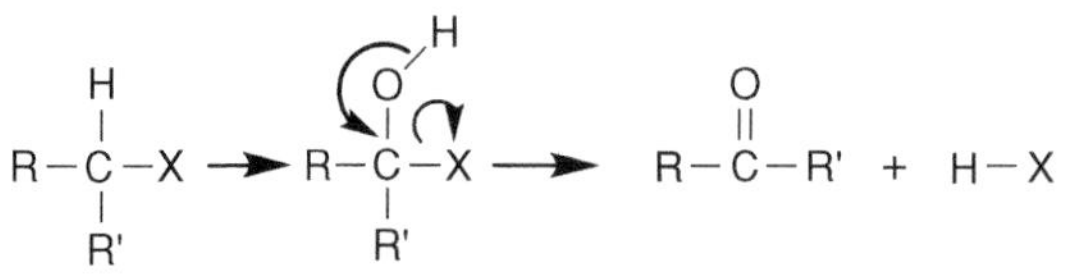

FIGURE 4.61 General scheme for oxidative dehalogenation.

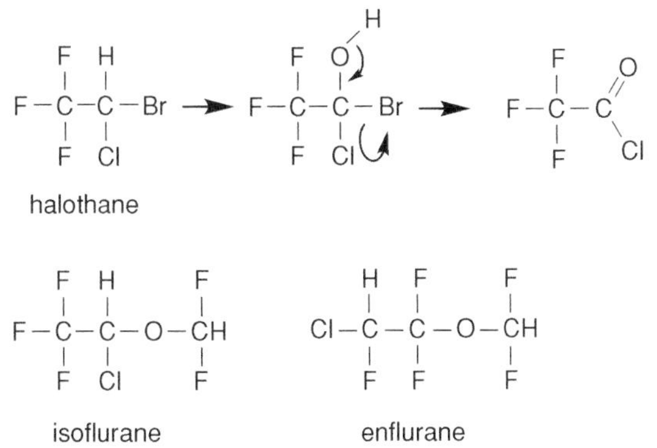

halothane

isoflurane   enflurane

FIGURE 4.62   Oxidative dehalogenation of halothane to form a reactive acid chloride intermediate and structures of other anesthetics that can form similar reactive metabolites.

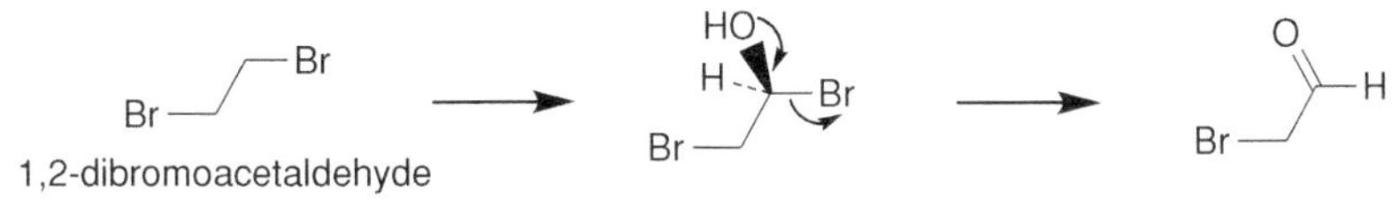

1,2-dibromoacetaldehyde

FIGURE 4.63   P450-catalyzed oxidation of 1,2-dibromoethane to bromoacetadehyde.

are being developed as a potential replacement for the chlorofluorohydrocarbons (CFCs) that are used as refrigerants, propellants, and dry-cleaning agents. The CFCs have been implicated as a cause of stratospheric ozone depletion, a destructive property that the HCFCs are designed to avoid by rendering them more biodegradable through the structural inclusion of a carbon–hydrogen bond. Unfortunately, the structural property that would make them useful replacement agents is also the same property that potentially makes them toxic to humans upon exposure. If the HCFC is capable of forming an acyl halide after carbon–hydrogen bond hydroxylation, e.g., the target carbon bears two halogens in addition to hydrogen, it is likely to be toxic. For example, studies indicate that its metabolic profile and ability to form metabolites like trifluoroacetyl chloride that covalently bind to protein mirror the toxicity of halothane (128). In contrast, HCFCs like HCFC-132b, HCFC-133a, HCFC-141b, which do not form acyl halide intermediates but rather aldehydes, show no indication of protein adduction or toxicity (129) (Fig. 4.64).

To probe the effects of HCFC structure on toxicity the metabolism of three penta-haloethanes, HCFC-123, HCFC-124, and HCFC-125 were studied. The three compounds differ one from the other by the number of fluorine atoms present in the β-carbon (Fig. 4.64). It was found that the enthalpies of activation, ΔH_{act}, for hydrogen atom abstraction paralleled the rate of trifluoroacetic acid excretion suggesting that the more difficult it was

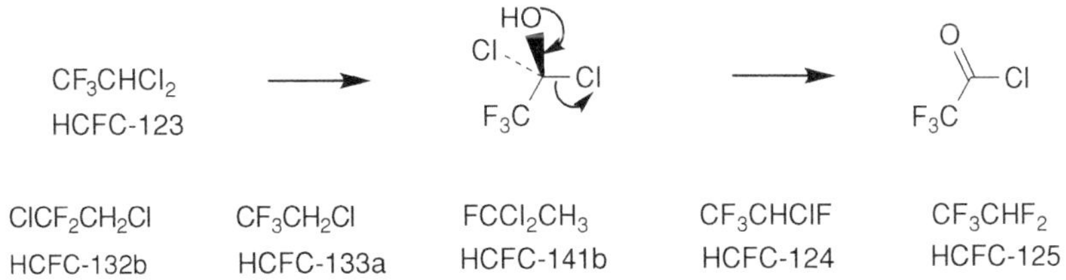

CF_3CHCl_2
HCFC-123

$ClCF_2CH_2Cl$	CF_3CH_2Cl	$FCCl_2CH_3$	CF_3CHClF	CF_3CHF_2
HCFC-132b	HCFC-133a	HCFC-141b	HCFC-124	HCFC-125

FIGURE 4.64   P450-catalyzed oxidation of HCFC-123 to trifluoroacetyl fluoride and the structures of other HCFCs.

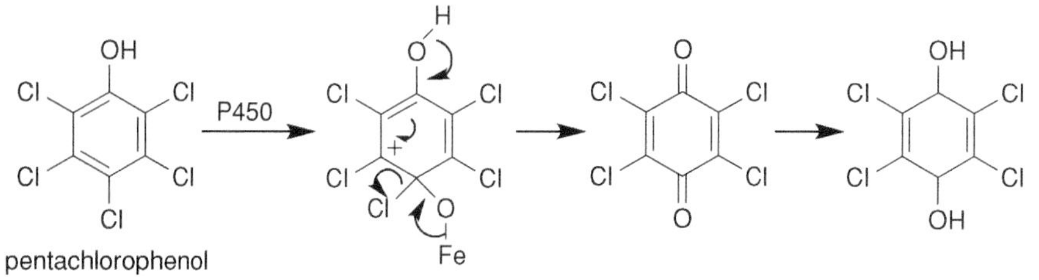

FIGURE 4.65 Mechanism of oxidative dehalogenation of an aryl halide.

to remove the hydrogen from the β-carbon, e.g. HCFC-125 vs. HCFC-123, the slower the rate of acyl halide formation (implied by the slower rate of appearance of trifluoroacetic acid) and presumably the lower the potential for toxicity. When the study was expanded to include 19 HCFCs, an excellent linear correlation was found between ΔH_{act} and the in vitro microsomal (primarily CYP2E1) or expressed human CYP2E1 rates of metabolite formation (130). These data strongly suggest that the likelihood of toxicity due to acyl halide formation in this class of substrates is a predictable phenomenon.

Oxidative dehalogenation of aromatic halogens should not occur because there is no hydrogen atom on the carbon involved; however, it often does occur. One mechanism likely involves ipso addition as will be discussed later and as proposed for the dechlorination of pentachlorophenol (Fig. 4.65) (131).

Halogen can also be removed either reductively, as will be discussed later in Chapter 5, or by glutathione displacement (Chapters 7 and 8) and as such represents a chemical group that is fairly labile in a biological environment.

Oxidative Cleavage of Esters and Amides

By analogy to N- and O-dealkylation reactions, one might expect esters and amides to be susceptible to P450-catalyzed oxidative attack at the α-carbon to oxygen (esters) or α to nitrogen (amides). This is indeed the case and was first established (132) by demonstration that the pyridine diester (Fig. 4.66) was oxidatively cleaved by rat-liver microsomes to yield the monoacid as shown.

Subsequently, it was shown (133) that P450 could catalyze the oxidative cleavage of a series of simple esters and several amides. These results suggest that oxidative cleavage is a general reaction for cytochrome P450s and commonly used esters and amides. But it is a reaction that has not been generally recognized because of being obscured by hydrolysis, especially in the case of esters. Unless the aldehyde product derived from the oxidative reaction was specifically sought and detected, the natural assumption for the breakup of an ester would be that it was a result of hydrolysis. Moreover, the aldehyde formed is usually

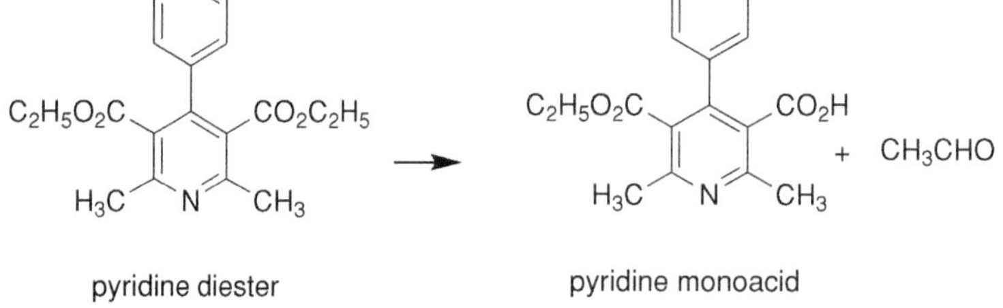

FIGURE 4.66 P450-catalyzed oxidative O-dealkylation of an ester.

oxidized to a carboxylic acid and the same products can be formed by oxidation of the alcohol formed by hydrolysis.

Oxidative Cleavage of Nitriles

Cytochrome P450-catalyzed carbon–carbon bond cleavage is a relatively rare event, but a simple case is the P450-catalyzed conversion of a nitrile to cyanide ion and an aldehyde or ketone depending on whether the α-carbon bears one or two hydrogen atoms. A simple example is acetonitrile as shown in Figure 4.67. The reaction mechanism is identical to N-dealkylation except the α-carbon is adjacent to the carbon of a cyano group rather than the nitrogen of an amino group. Both cyano and protonated alkyl nitrogen are effective leaving groups. The reaction has been modeled (134) using a set of 26 structurally diverse nitriles, and it was found that acute toxicity in the mouse correlated with the ease of hydrogen abstraction α to the nitrile that leads to cyanide release, relative to oxidative attack at other intramolecular sites that might lead to elimination without cyanide release.

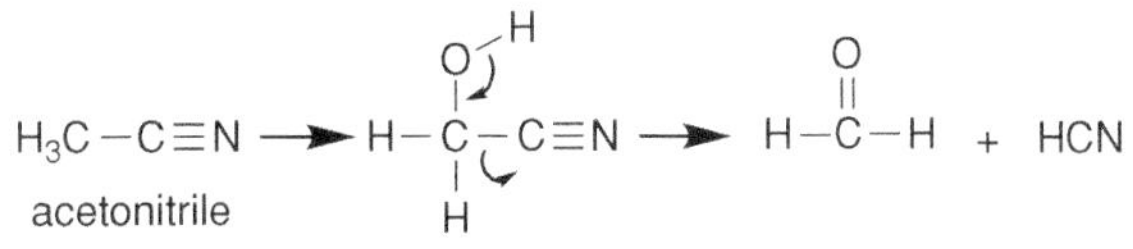

FIGURE 4.67   Oxidative cleavage of acetonitrile.

Oxidative Dehydrogenation

Oxidation of a saturated hydrocarbon almost invariably leads to the formation of an alcohol except for a few cases where unsaturation is introduced. After initial abstraction of a hydrogen atom to generate a carbon-based radical, a competition between oxygen rebound versus abstraction of a second hydrogen atom from an adjacent carbon becomes operative and product formation is partitioned between alcohol and alkene. An example can be found in the metabolism of the anticonvulsant valproic acid. The drug undergoes normal cytochrome P450-catalyzed transformation to generate the ω and $\omega - 1$ hydroxylated metabolites, respectively. However, the unsaturated substrate, 4-ene valproic acid, is also produced (Fig. 4.68). While the 4-ene is a minor metabolic product, its formation is important both from a mechanistic perspective and from the fact that it is toxic. Deuterated valproic acid analogs were used to establish (135) that the 4-ene metabolite forms in competition with 3-hydroxy metabolite after initial abstraction of a hydrogen atom from the $\omega - 1$ carbon.

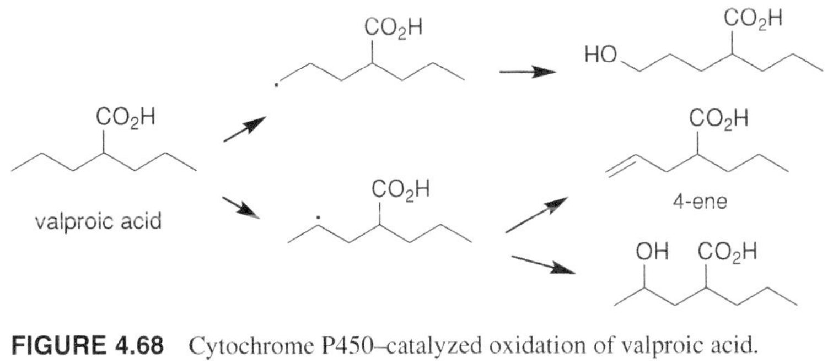

FIGURE 4.68   Cytochrome P450–catalyzed oxidation of valproic acid.

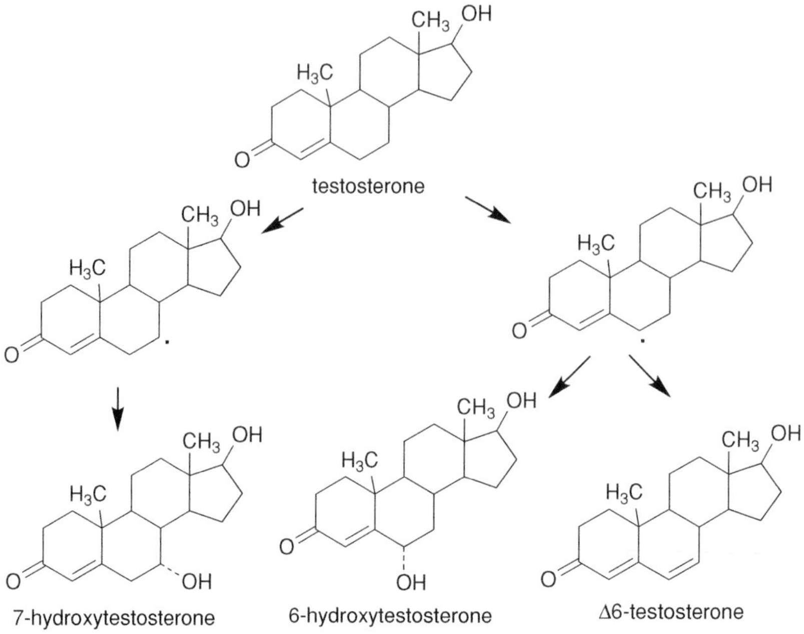

FIGURE 4.69 Cytochrome P450–catalyzed oxidation of testosterone.

A second example can be found in the metabolism of the male sex steroid hormone testosterone. In addition to 7α-hydroxy- and 6α-hydroxytestosterone, CYP2A1 was also found to form Δ6-testosterone (136) (Fig. 4.69). Using selectively deuterated analogs (136), Δ6-testosterone was established as being formed in competition with 6α-hydroxytestosterone after initial hydrogen atom abstraction from C6 to form the common radical intermediate. Little, if any, Δ6-testosterone was found to be formed in competition with 7α-hydroxytestosterone after initial hydrogen abstraction from C7.

A final example of dehydrogenation to form an alkene is ezlopitant, a drug being developed as a potential substance P receptor antagonist (Fig. 4.70). Metabolism of ezlopitant results in both the benzyl alcohol and the corresponding alkene being found as major metabolites. Interestingly, the alkene does not arise from dehydration of the benzyl alcohol. Rather selectively deuterated analogs of ezlopitant revealed that the alkene, similar to the first two examples, arose in competition with benzyl alcohol formation after initial abstraction of the benzylic hydrogen (137).

The most common examples of oxidative dehydrogenation involve a carbon–heteroatom bond such as the oxidation of acetaminophen, methylformamide, and 2-hydroxycarbamazepine (Fig. 4.71) (79).

Oxidative Addition to Unsaturated Carbon

Alkenes
The hybridization of the carbon in an alkene makes it even more difficult to break the carbon–hydrogen bond of a vinylic carbon than of a saturated carbon. As a consequence, cytochrome P450, rather than abstracting a hydrogen atom, catalyzes the addition of an oxygen atom to the double bond leading to the formation of an epoxide as shown in Figure 4.72.

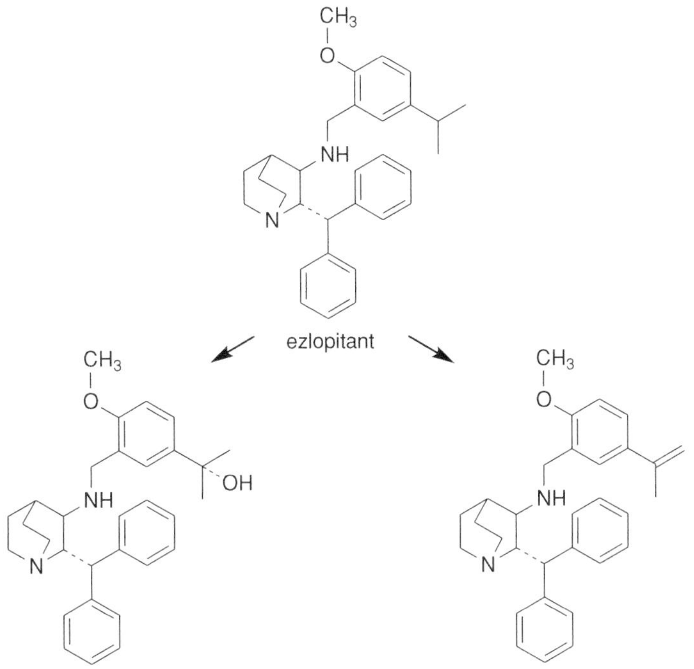

FIGURE 4.70 P450-catalyzed oxidation of ezlopitant.

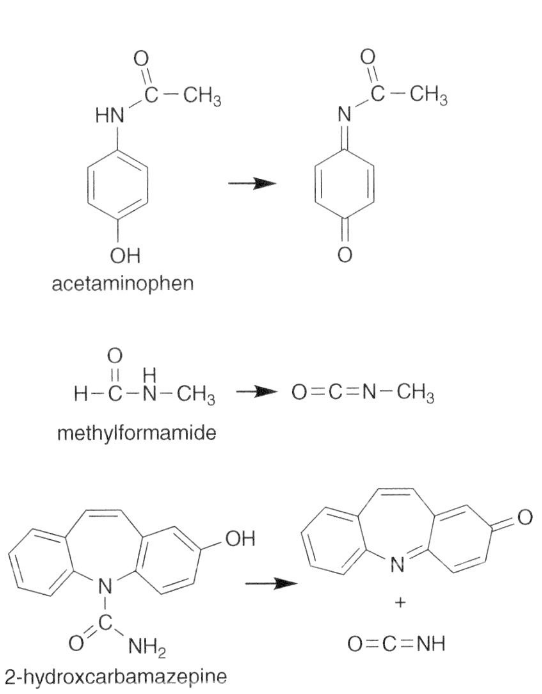

FIGURE 4.71 Examples of substrates that undergo oxidative dehydrogenation involving heteroatoms.

FIGURE 4.72 General scheme for the oxidation of an alkene to an epoxide.

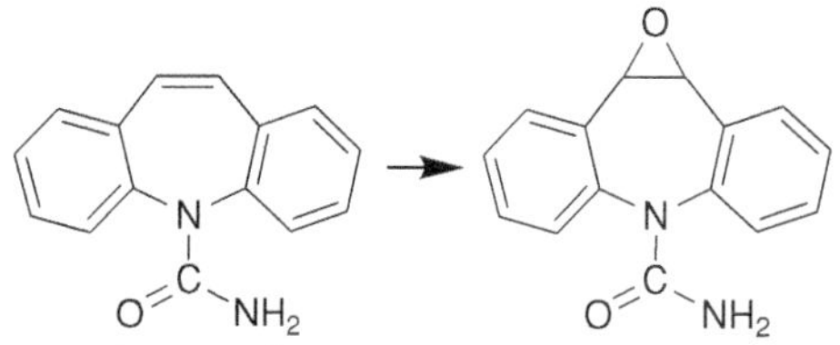

carbamazepine

FIGURE 4.73 Oxidation of carbamazepine to an epoxide.

The major metabolic pathway of carbamazepine is an example of this oxidation as shown in Figure 4.73. The usual bond angle of a sp^3-hybridized carbon is 109°, but it is constrained to be 60° in an epoxide making epoxides reactive. This reactivity varies significantly depending on the structure of the epoxide, and this will be discussed further in Chapter 8. The epoxide of carbamazepine is relatively unreactive and easy to isolate.

Mechanistically the reaction is bounded by two extremes. At one extreme, FeO^{3+} adds to the double bond in a single step. At the other, a two-step reaction involving the generation of an intermediate is operative. The path taken goes to the very nature of the active oxygen. If it is singlet-like, one might expect a synchronous insertion of the oxygen atom characterized by a single transition state since spin inversion would not be required to complete formation of the two new carbon–oxygen bonds. If it is triplet-like, then spin inversion would be required, after formation of the first carbon–oxygen bond and generation of an adjacent carbon radical, to allow oxygen rebound to form the second new carbon–oxygen bond. Early studies had indicated that epoxidation proceeds with retention of configuration, which is supportive of a concerted mechanism involving singlet-like oxygen. Subsequent studies found that P450-catalyzed epoxidation of a number of terminal mono-substituted olefins, e.g., ethylene, propene, 1-octene, vinyl fluoride, was accompanied by a competing oxidative pathway consistent with a nonconcerted multiple-step mechanism that led to suicide destruction of the enzyme by selectively alkylating the pyrrole nitrogen of ring D (138). Like the hydroxylation of simple hydrocarbons, the data appears to be paradoxical with regard to mechanism. But again the two-state reactivity paradigm offers a theoretical model that resolves the dilemma (139). The epoxide with conserved stereochemistry results primarily from the low-spin doublet state, while the high-spin quartet state leads to formation of the epoxide by a two-step mechanism.

Alkynes

Cytochrome P450-catalyzed oxidation of terminal aryl alkynes generates the corresponding substituted aryl acetic acid. To investigate the reaction, the alkyne hydrogen of 4-ethynylbiphenyl was replaced with deuterium and was found to be quantitatively retained on the α-carbon of the acid metabolite (140) (Fig. 4.74). Addition of FeO^{3+} to the terminal carbon of the acetylene group with concerted migration of the terminal acetylenic hydrogen to the adjacent carbon leads to the formation of a substituted ketene. Hydrolysis of the ketene generates the aryl acetic acid as final product. If FeO^{3+} adds to the inner carbon of the acetylene group rather than the terminal carbon, the reaction takes an entirely different course. This reaction pathway leads to alkylation of a heme nitrogen and

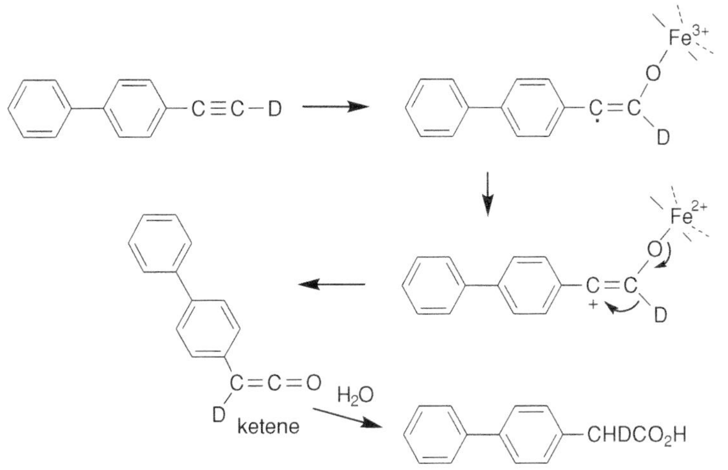

FIGURE 4.74   Mechanism for the oxidation of terminal aryl alkynes.

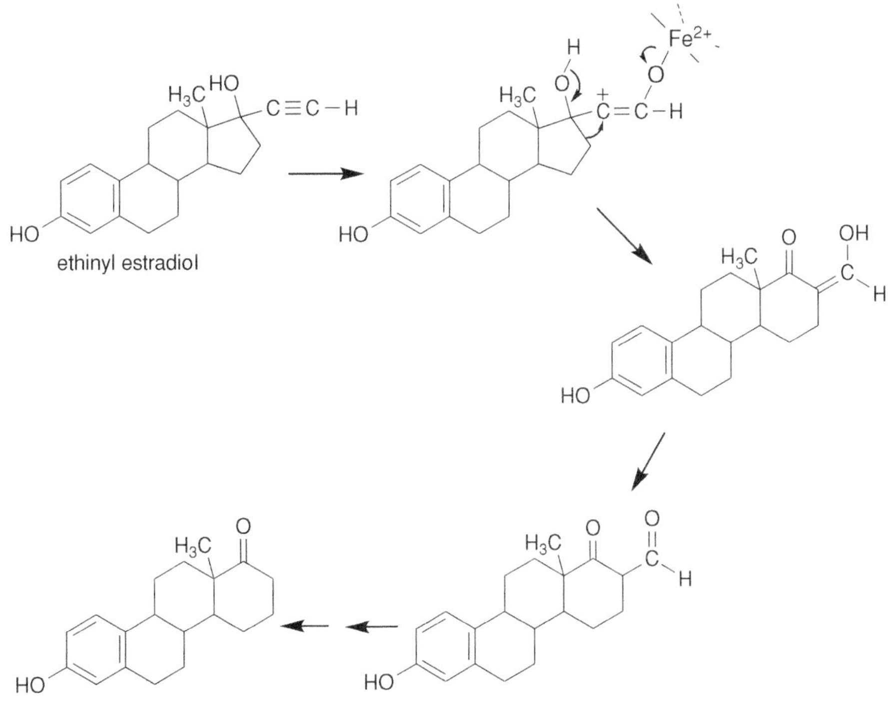

FIGURE 4.75   Oxidation of the acetylenic group of ethinyl estradiol with rearrangement leading to ring expansion.

destruction of the enzyme. As a bond between FeO^{3+} and the inner carbon begins to form, one of the heme nitrogen atoms will tend to complex with the terminal carbon to stabilize the developing electron deficiency. The end result is the formation of covalent bond between the heme nitrogen and the terminal carbon (141).

There are not many drugs that are alkynes; however, one good example is ethinyl estradiol (Fig. 4.6). Even though ethinyl estradiol is not an aryl alkyne, the acetylenic group is attached to a tertiary carbon and not adjacent to an α-carbon–hydrogen bond.

Thus, it would be reasonable to expect that the acetylenic group would be metabolized to a carboxylic acid. While the acid metabolite has not been detected, ethinyl estradiol is a suicide substrate inhibitor of CYP2B6 consistent with a reactive ketene being formed as an intermediate (142). In addition, there is a rearrangement product (4) shown in Figure 4.75.

Aromatic Rings

The frequency with which aromatic hydroxylation is found as a metabolic event is undoubtedly a reflection of the commonality of an aromatic ring(s) as a structural component(s) of most drug molecules. While the hydroxylated product of this pathway is usually not the major metabolite of a given aromatic ring–containing drug, it is often found as a significant contributor to the overall metabolic profile of that drug. In general, regioselectivity of P450-catalyzed aromatic hydroxylation follows the rules of electrophilic aromatic hydroxylation established by the linear free-energy relationships of physical organic chemistry, i.e., para > ortho > meta in most systems. This order prevails unless the system is deactivated by a substituent that on balance withdraws electron density from the ring, e.g., nitro group, in which case meta-substitution dominates since it is the site that is the least deactivated toward electrophilic attack. In the case of P450 catalysis, an exception would occur if the steric demands of the active site architecture of the enzyme for a specific substrate favored meta-hydroxylation. These general observations suggest that if the enzyme has a sterically permissive active site that is not overly restrictive to substrate motion, the electronic properties of the substrate should determine the regioselectivity of hydroxylation. These insights have led to the development of computational models for predicting aliphatic hydroxylation, aromatic hydroxylation, or a combination of both pathways (109,115,130). The models are not only promising in their predictive capacity but have already met with considerable success.

That cytochrome P450-catalyzed aromatic hydroxylation proceeded by a mechanistic pathway that was generally consistent with the rules of electrophilic aromatic substitution was never in doubt because of the abundance of experimental evidence supporting this conclusion. Despite the certainty of product formation, establishing the exact mechanism that defines the pathway has proved to be difficult.

One of the first pieces of evidence for the mechanism of this reaction involved an attempt to develop a new assay for the activity of tyrosine synthase, which converts phenylalanine to tyrosine. A tritium was placed in the para position of phenylalanine, and it was assumed that oxidation of this position would lead to the loss of tritium and the rate of this loss would be a measure of the activity of the enzyme (Fig. 4.76).

However, when the results of this assay were compared to other assays, it was found to underestimate the activity of the enzyme. Further analysis revealed that some of the tyrosine contained tritium in the meta position; this was referred to as the NIH shift because the early mechanistic studies were performed at the National Institutes of Health (143). The

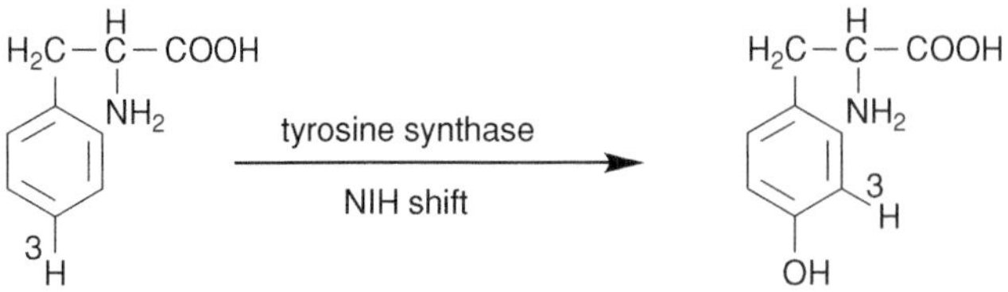

FIGURE 4.76 Cytochrome P450–catalyzed oxidation of *p*-tritiated phenylalanine.

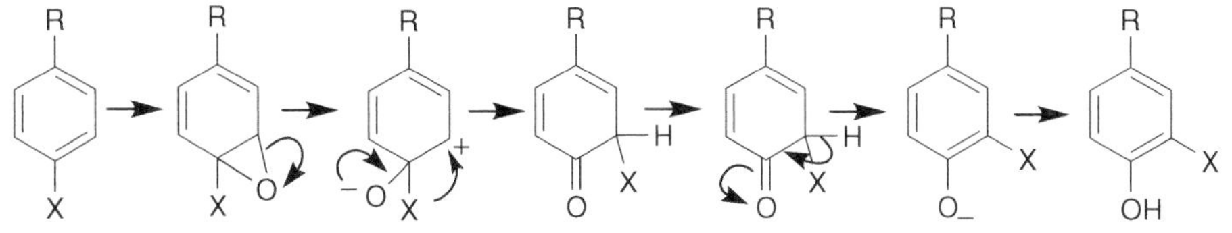

FIGURE 4.77   The original proposal for the mechanism of the NIH shift.

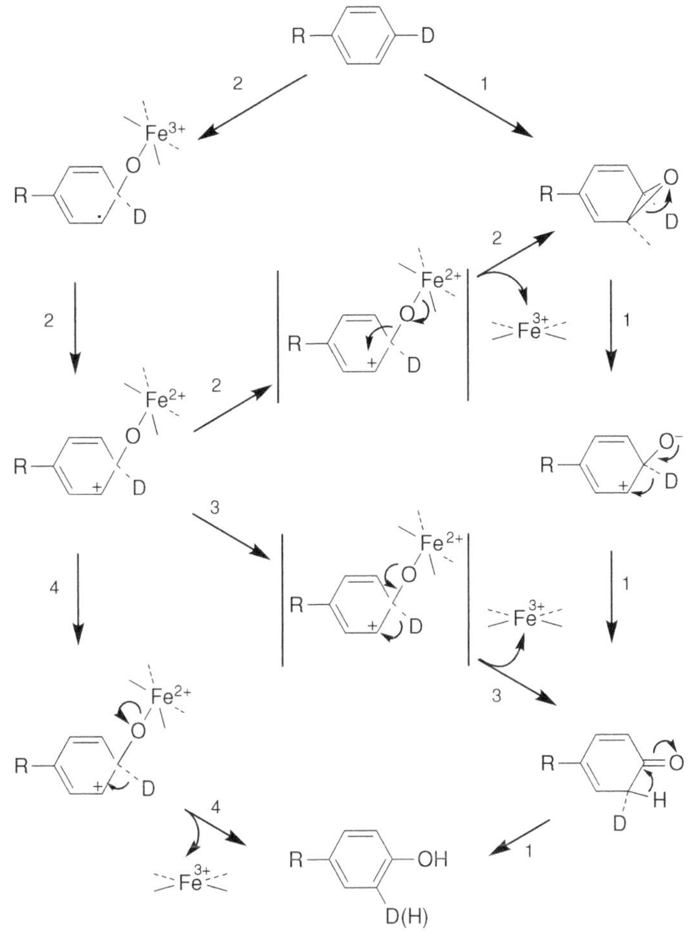

FIGURE 4.78   Mechanistic pathways for aromatic hydroxylation by concerted addition of oxene, pathway 1, or by stepwise addition of oxene, pathway 2. Pathways 2, 3, and 4 describe the formation of phenol that bypasses the arene oxide intermediate.

originally proposed mechanism for the NIH shift is shown in Figure 4.77, but it is probably an oversimplification of the true state of affairs.

As with an isolated double bond, epoxide formation in an aromatic ring, i.e., arene oxide formation, can occur mechanistically either by a concerted addition of oxene to form the arene oxide in a single step, pathway 1, or by a stepwise process, pathway 2 (Fig. 4.78). The stepwise process, pathway 2, would involve the initial addition of enzyme-bound FeO^{3+} to a specific carbon to form a tetrahedral intermediate, electron transfer from the aryl group to heme to form a carbonium ion adjacent to the oxygen adduct followed by

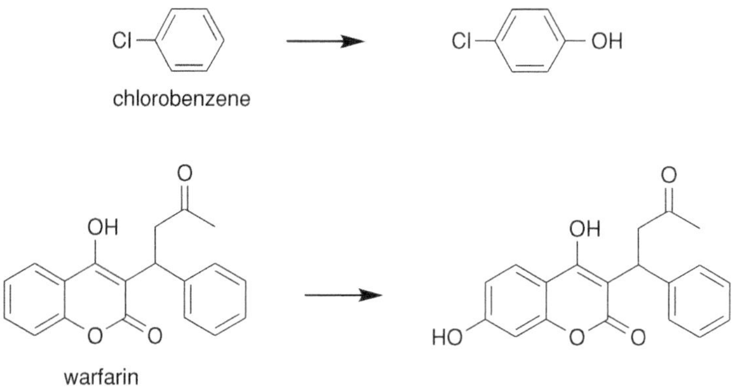

FIGURE 4.79 Substrates, chlorobenzene and warfarin, used to test whether phenol formation involves a stepwise mechanism.

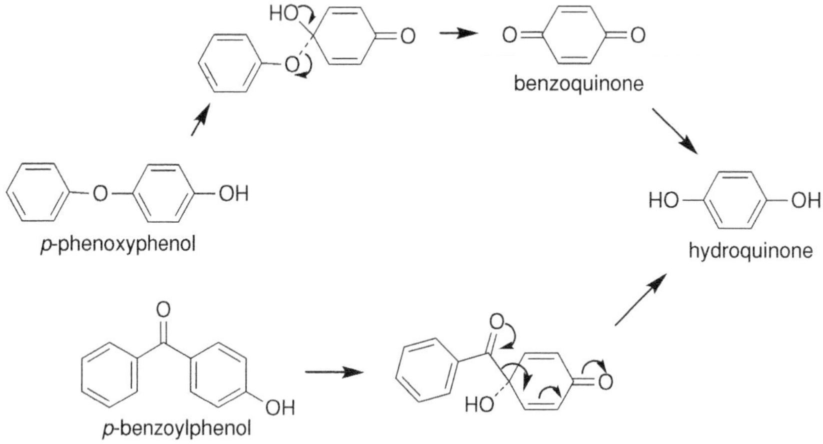

FIGURE 4.80 Oxidative mechanisms for the formation of hydroquinone from *p*-substituted phenols.

ring closure to the arene oxide. From this point on to the final phenol product formation, both mechanisms are identical. The ring opens generating an adjacent carbonium ion. Hydride shifts to satisfy the adjacent carbonium ion as a pair of electrons from oxygen move in to satisfy the positive charge being developed as hydride leaves. The overall process leads to ketone formation that then tautomerizes to generate the phenol. The stepwise mechanism does not necessarily have to close to the arene oxide. At the early stage of carbonium ion formation (pathway 2), it can bypass arene oxide formation to form the ketone directly (pathway 3), then proceed on to phenol, or it can even bypass ketone and form the phenol directly (pathway 4) (Fig. 4.78). A concerted mechanism would indicate that an arene oxide is an obligatory intermediate on the path to phenol whereas a stepwise mechanism, as indicated, would not necessarily have to pass through the epoxide.

A theoretical study (144) and several experimental studies using selectively deuterated mono-substituted benzenes (145), chlorobenzene (146), and warfarin (147) (Fig. 4.79) provided strong evidence for the stepwise mechanism. Most recently, a theoretical study using density functional calculations reported the same basic conclusion (148). The reaction proceeds by a stepwise mechanism involving initial attack of FeO^{3+} on the π system to form a tetrahedral intermediate (pathway 2). The tetrahedral intermediate then goes on to ultimately form phenol, either directly (proton transfer to a pyrrole nitrogen from the tetrahedral carbon) (pathway 4) or indirectly via ketone (pathway 3) or arene oxide (pathway 2) (Fig. 4.78).

It is sometimes assumed that every phenol metabolite indicates the formation of an arene oxide intermediate; however, as discussed above, arene oxides are not obligate intermediates in the formation of phenols. This is an important distinction because arene oxides and other epoxides are reactive intermediates that can be toxic or even carcinogenic, e.g., epoxides of some polycyclic aromatic hydrocarbons. The question of whether their formation is obligatory is significant for drug design and development and has implications for toxicity as discussed in Chapter 8.

Ipso Substitution

A recently recognized aspect of P450-catalyzed aromatic hydroxylation is the formation of hydroquinone from a para-substituted phenol, i.e., replacement of the para-substituent with a hydroxy group—a phenomenon termed ipso substitution. The reaction requires the phenolic group, but the scope of the reaction is fairly broad and not totally limited to good leaving groups such as halides. For example, nine p-substituted phenols (F, Cl, Br, NO_2, CN, CH_2OH, $COCH_3$, COPh, and CO_2H) of diverse structure were incubated with rat-liver microsomes and the amount of hydroquinone formed from each substrate was determined (149). After para addition of the oxidant to form the tetrahedral semiquinone-like intermediate, elimination to form the final product can be accommodated by loss of either a negatively or positively charged substituent depending on the nature of the substituent (Fig. 4.80). For example, in the case of p-phenoxyphenol, phenol is lost as the anion and leads to formation of benzoquinone, which is subject to rapid reduction to hydroquinone. Groups such as CH_2OH, $COCH_3$, or p-benzoyl are lost as the corresponding cations producing hydroquinone directly, e.g., the loss of the benzoyl group from p-benzoylphenol.

FeO_2^+—An Active P450 Oxidant

The mechanism of cytochrome P450-catalyzed oxidative reactions discussed thus far has focused on the ability of the active oxidant, oxene, to initiate reaction in one of two different ways—either by abstracting a hydrogen atom from a carbon hydrogen bond or in the case of aromatic systems, adding to the π system. However, oxene is not the only active oxidizing species formed by P450. The peroxy anion species, FeO_2^+, generated upon addition of the second electron to enzyme-bound molecular oxygen is also an active oxidizing species, although nucleophilic in character. If a suitable substrate is bound to the enzyme when the peroxy anion is generated, it can react with the substrate in competition with protonation that leads to the loss of water and the generation of oxene, FeO^{3+}. Suitable substrates would appear to be ones that are susceptible to nucleophilic attack by the peroxy anion, e.g., aldehydes.

If cyclohexanecarboxaldehyde is incubated with CYP2B4, NADPH, and cytochrome P450 reductase, the aldehyde–cyclohexyl ring carbon–carbon bond is cleaved generating cyclohexene and formic acid (150) (Fig. 4.81). The reaction is supported if hydrogen peroxide replaces NADPH and cytochrome P450 reductase but is not supported if other oxidants at the same oxidation equivalent as peroxide but bypass the peroxy form of P450 such as iodosobenzene, m-chloroperbenzoic acid, or cumyl hydroperoxide are used. These

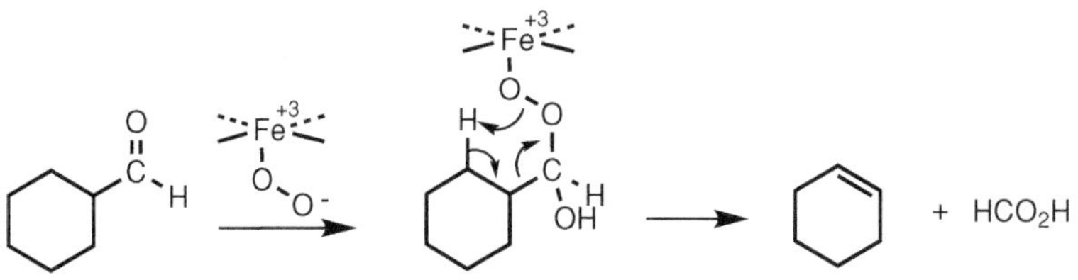

FIGURE 4.81 Cytochrome P450–catalyzed oxidation of cyclohexylcarboxaldehyde.

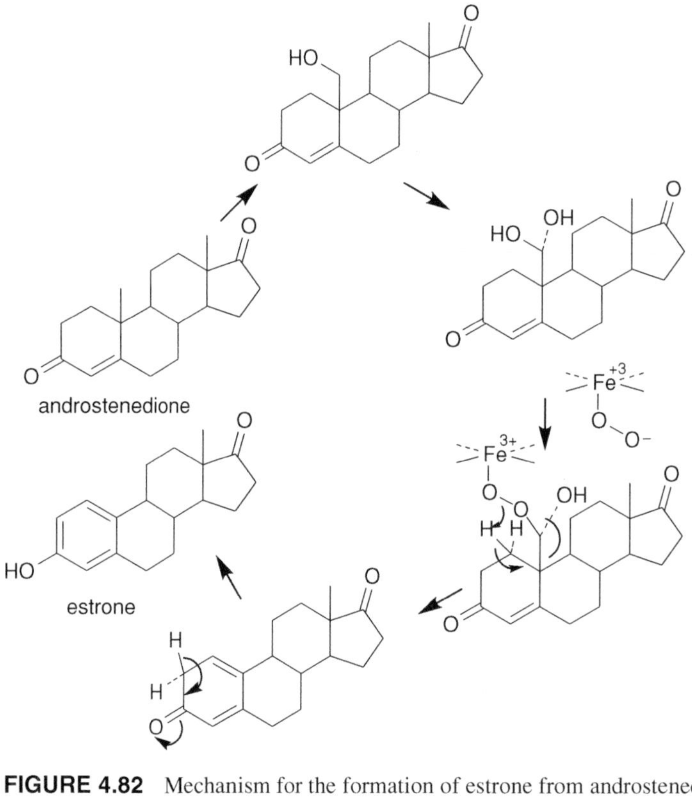

FIGURE 4.82 Mechanism for the formation of estrone from androstenedione by aromatase.

results suggest that an O_2-derived heme iron-bound peroxide attacks the carbonyl carbon to form an enzyme-bound peroxyhemiacetal-like intermediate. The intermediate rearranges either by a concerted or sequential mechanism to yield the observed products (150).

The P450-catalyzed oxidation of cyclohexanecarboxaldehyde was initially studied as a model for the final step in the conversion of androstenedione to estrone by the steroidogenic P450, aromatase, through a remarkable sequence of reactions that leads to the selective cleavage of an unactivated carbon–carbon bond (Fig. 4.82). The first two steps of the reaction sequence involving initial hydroxylation of C19 followed by a second hydroxylation of C19 to form a hydrated aldehyde were well understood and characterized. But the final oxidation leading to loss of formic acid and aromatization of the A-ring was not, and the cyclohexanecarboxaldehyde model reaction helped to establish and clarify the likely mechanism. The scope of the CYP2B4-catalyzed reaction appears to accommodate the selective deformylation of a number of simple α- or β-branched chain, but not normal chain, aldehydes to generate alkenes (151).

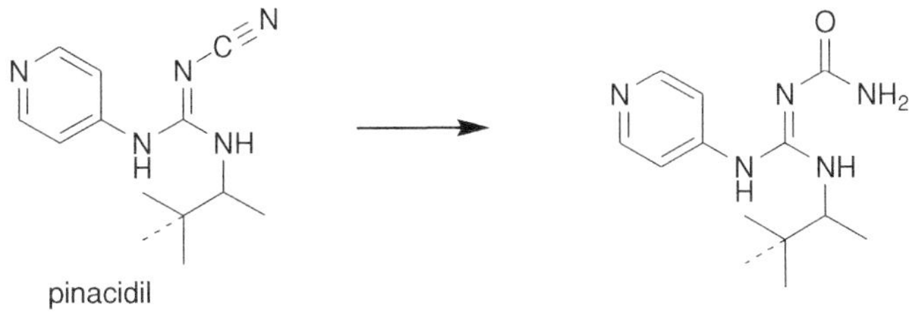

pinacidil

FIGURE 4.83 P450-catalyzed oxidative conversion of the cyano group of pinacidil to the amide.

While the cyanide group of nitriles can be oxidatively removed by P450 catalysis, they can also be oxidized to amides. It has been found that CYP3A4 converts the cyano group of potassium channel–opening agent, pinacidil, to the corresponding amide (152) (Fig. 4.83). Moreover, the amide was also obtained if hydrogen peroxide replaced NADPH and P450 reductase in the reaction suggesting that the active oxidant was probably the nucleophilic FeO_2^+ species rather than FeO^{3+}.

Oxidation of Alcohols

Oxidation of primary alcohols leads to aldehydes and oxidation of secondary alcohols leads to ketones. This oxidation also involves the loss of two hydrogen atoms. However, unlike the oxidations discussed so far in this chapter that are mediated almost exclusively by cytochromes P450, the major enzyme involved in the oxidation of ethanol is ALD (discussed earlier in this chapter) (74). Although ALD is the major enzyme involved in the oxidation of ethanol and most other low molecular-mass alcohols, cytochromes P450, especially 2E1, can also oxidize ethanol and this enzyme is induced in alcoholics. Although comprehensive studies have not been published, it appears that cytochromes P450 are often the major enzymes involved in the oxidation of higher molecular mass alcohols.

Oxidation of Aldehydes

Aldehydes are oxidized to carboxylic acids. A major enzyme responsible for this oxidation is aldehyde dehydrogenase (see "Aldehyde Dehydrogenases" section in this chapter) (79); however, other enzymes such as AO and cytochromes P450 can also mediate the oxidation of aldehydes as discussed (discussed earlier in this chapter). Ketones are not substrates for aldehyde dehydrogenase for the same reason that tertiary alcohols cannot be oxidized by ALD. Unlike the oxidation of alcohols, the oxidation of aldehydes is irreversible. Aldehydes are usually toxic and therefore there are aldehyde dehydrogenases in virtually all cells and in most compartments within cells.

Oxidative Decarboxylation

Metabolic pathways rarely lead to breaking a carbon–carbon bond; however, there are exceptions such as the conversion of the prodrug nabumetone to an active nonsteroidal anti-inflammatory agent as shown in Figure 4.84 (153). Although the mechanism of this conversion is unknown, if oxidation leads to two adjacent carbonyl groups it weakens the carbon–carbon bond and further oxidation leads the rupture of this bond.

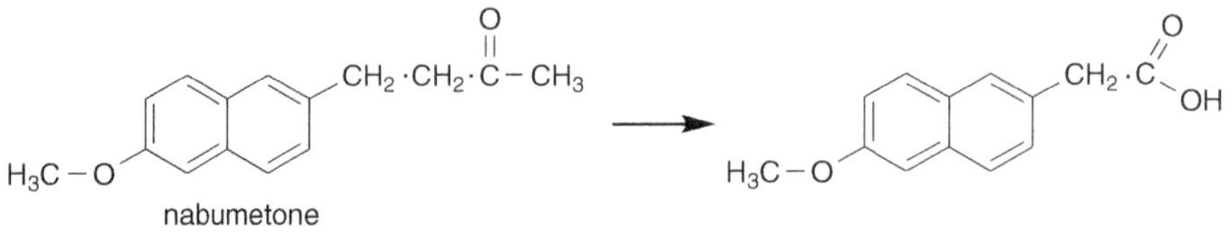

FIGURE 4.84 Example of an unusual metabolic pathway leading to breaking a carbon–carbon bond.

Oxidation of Heteroatoms

Oxidation of Nitrogen

Cytochrome P450-catalyzed oxidation of heteroatom-containing drugs commonly leads to dealkylation of the heteroatom. As we have seen, N-dealkylation is among the most frequently observed and one of the least energetically demanding metabolic reactions. But a second pathway is possible. It involves direct addition of oxene to the heteroatom to form N-oxides, in the case of tertiary amines, or hydroxylamines in the case of primary or secondary amines. As indicated earlier, N-dealkylation is thought to proceed by either the SET pathway, i.e., initial electron abstraction from nitrogen, or the HAT pathway, i.e., initial abstraction of an α-hydrogen atom (Fig. 4.51). If the SET pathway is operative, then generation of a nitrogen-based radical cation could serve as a common intermediate for either N-dealkylation or the addition of oxene. The result would be either N-hydroxylation or N-oxide formation. Which metabolite pathway prevails would then be a matter of competition between the rate of proton loss from the α-carbon versus the rate of oxygen radical rebound to nitrogen. The HAT pathway requires N-dealkylation and oxidative attack at nitrogen to be totally independent processes. Evidence that definitively establishes one of these pathways to the total exclusion of the other has not been presented. A third possibility is that the energetics for initial electron abstraction versus hydrogen atom abstraction are not that far apart and that both pathways are possible. The one that prevails for any given substrate depends on substrate structure.

The first product in the oxidation of primary amines is a hydroxylamine as indicated in Figure 4.85. Hydroxylamines can be further oxidized to nitroso metabolites, which can be viewed as analogous to the oxidation of an alcohol to an aldehyde. If there is a hydrogen

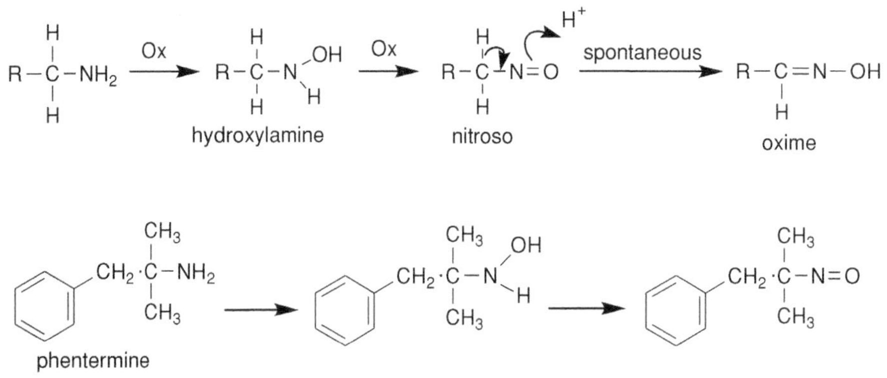

FIGURE 4.85 Oxidation of primary amines leads to a hydroxylamine followed by a nitroso metabolite, which if there is a hydrogen on the α-carbon can rearrange to an oxime. Without such a hydrogen, as in the case of phentermine, no rearrangement is possible.

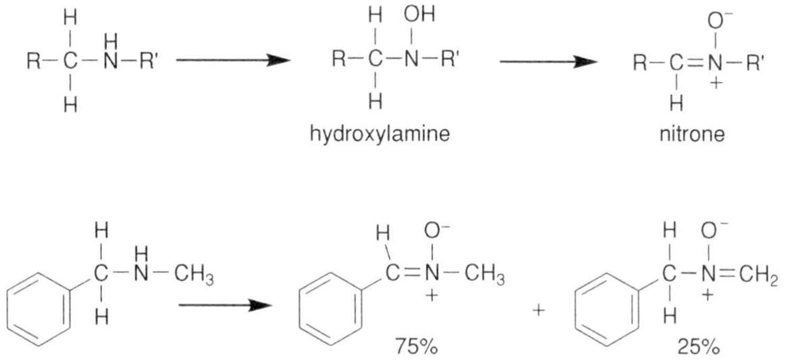

FIGURE 4.86   Oxidation of a secondary amine to a hydroxylamine followed by nitrone formation.

on the α-carbon, nitroso metabolites usually spontaneously rearrange to oximes; without such an α-hydrogen such a rearrangement cannot occur as in the case of phentermine.

The first product in the oxidation of secondary amines is also a hydroxylamine (Fig. 4.86). Further oxidation requires the involvement of an adjacent carbon atom to form a nitrone. If there is more than one adjacent carbon with a hydrogen atom, the major product will usually involve the most substituted carbon atom (assuming at least one hydrogen such that a new bond can be formed) or one in which the nitrone is conjugated with other double bonds as shown in Figure 4.86.

N-hydroxylation is not restricted to primary and secondary amines. For example, nitrogen-based functional groups such as amides, amidino, guanidino, hydrazino, etc. that have at least one nitrogen–hydrogen bond are susceptible to N-hydroxylation.

Amides that undergo N-hydroxylation are often amides of arylamines (Fig. 4.87) some of which are carcinogens such as 2-acetylaminofluorene (Fig. 4.87). Initial N-hydroxylation of a hydrazine is similar to that of an amine; however, further oxidation can lead to the formation of nitrogen gas and reactive species. The two-electron oxidation pathway is shown in Figure 4.87. Hydrazines also undergo one-electron oxidations but the intermediates are short-lived and these pathways are less well defined (154).

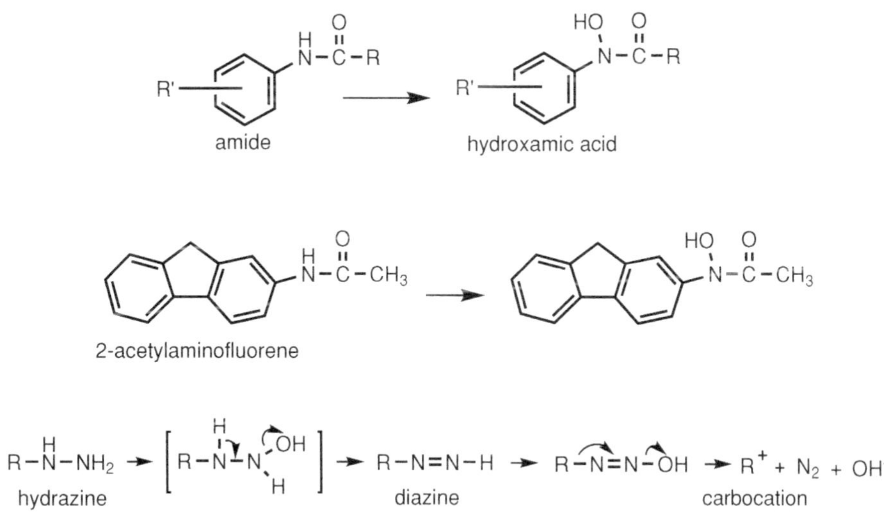

FIGURE 4.87   N-oxidation of amides and hydrazines.

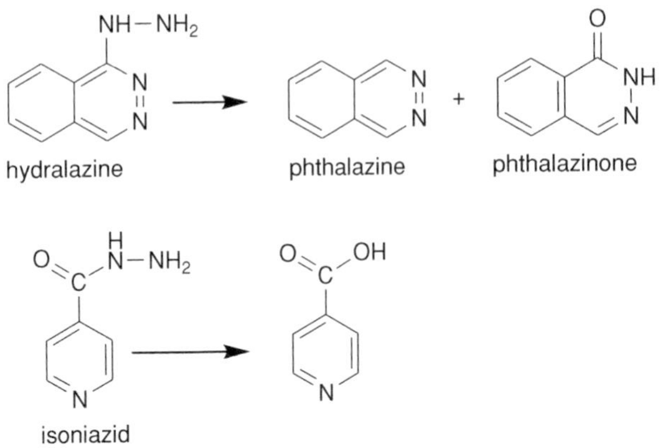

FIGURE 4.88 Oxidation of the hydrazine (hydralazine) and the hydrazide (isoniazid) leads to the loss of nitrogen.

There are few drugs that contain a hydrazine group and, in general, such drugs are associated with a high incidence of adverse reactions. Probably the least toxic is hydralazine, which is associated with a high incidence of drug-induced lupus. It is oxidized to phthalazine, possibly through a free-radical pathway, and phthalazinone (Fig. 4.88) probably through a carbocation, which can be trapped by N-acetylcysteine (155). Isoniazid is a hydrazide and is associated with a relatively high incidence of liver toxicity. It is oxidized to isonicotinic acid (156).

Tertiary amines cannot form hydroxylamines and the oxidation involves the nitrogen lone pair of electrons leading to an amine oxide, which has a coordinate covalent bond between the nitrogen and oxygen as shown in Figure 4.89. A pyridine-type nitrogen can also be oxidized to an N-oxide. There are many drugs that are tertiary amines; the example of imipramine is shown in Figure 4.89. The N-oxide is often pharmacologically inactive; however, N-oxides can be reduced back to the tertiary amine as will be discussed in Chapter 5 and therefore the N-oxide can act as a "depot" form of the drug.

It is clear from the literature that, while N-oxides and N-hydroxylated compounds are observable metabolites of amines, they are nowhere near as prevalent as metabolites resulting from N-dealkylation. They appear to only become more significant when N-dealkylation is not an option (157). A caveat of this conclusion however is that N-hydroxy compounds are not all that stable, particularly in vivo, where they can be reduced back to the amine or further oxidized to even less stable compounds. It may be that,

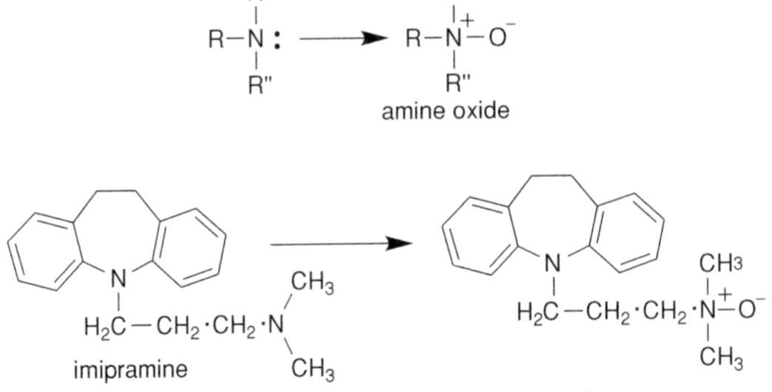

FIGURE 4.89 Oxidation of tertiary amines to N-oxides.

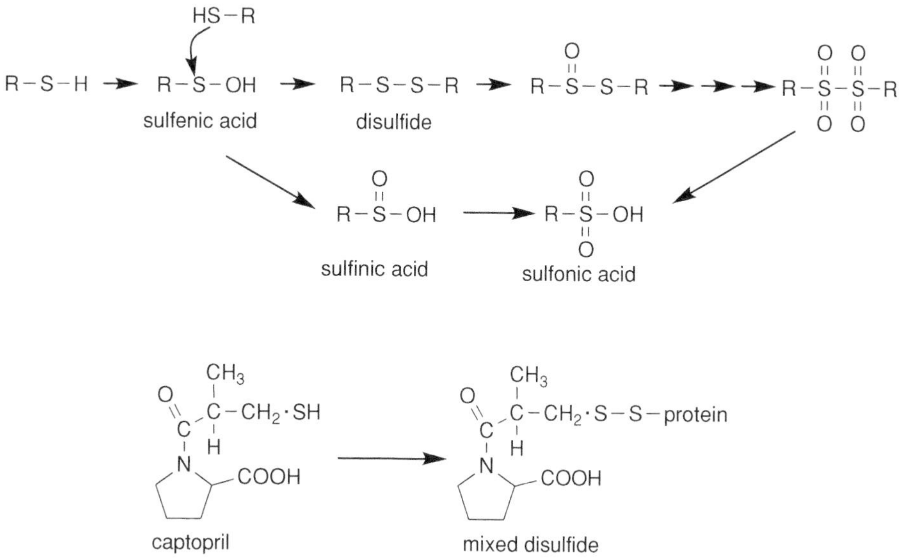

FIGURE 4.90 Oxidation pathways of thiols. In vivo the major product is a mixed disulfide as shown for captopril.

while they clearly are not major players, the degree to which they contribute to an overall metabolic profile might be underestimated because of their relative instability.

Oxidation of Sulfur

N-hydroxylation and N-oxide formation are minor pathways relative to N-dealkylation, but the exact opposite is true of sulfur oxidation. S-dealkylation is a minor pathway of metabolism, while direct oxidation of sulfur to form a sulfoxide and/or a sulfone is a major pathway. Even though the enzyme FMO can also catalyze sulfur oxidation, cytochrome P450 is often a major if not the major contributor. If P450-catalyzed sulfide oxidation is initiated, as seems likely, by abstraction of an electron to generate a sulfur radical cation that could serve as a common intermediate for sulfoxidation and S-dealkylation, product formation would reflect the competition between these pathways. Given that S-dealkylation is a minor pathway, proton loss from the α-carbon would need to be much slower than oxygen rebound. The expected decreased acidity of a carbon–hydrogen bond adjacent to a sulfur radical cation versus a nitrogen radical cation would be consistent with this requirement. However, independent mechanisms, a SET mechanism for sulfoxidation versus a HAT mechanism for S-dealkylation, could equally well account for the data since the energy required for ionization is much lower for sulfur than it is for nitrogen. Overall the results seem to suggest that the third mechanistic possibility prevails, i.e., both initial electron abstraction and hydrogen atom abstraction mechanisms are possible, but the one that dominates depends on the relative energetics for the two pathways.

Sulfur-containing compounds that are susceptible to oxidation include thiols, sulfides, and disulfides. In addition, some of their initial oxidation products can serve as substrates for further oxidation. For example, the oxidation of thiols (also known as sulfhydryl groups) leads to a sulfenic acid (Fig. 4.90). Sulfenic acids are reactive, the major reaction being reaction with other thiols to form disulfides (158). In vitro, this usually leads to disulfides of the drug; however, in vivo the concentration of biological thiols such as proteins or glutathione is higher than the drug, and therefore the major product is a mixed disulfide

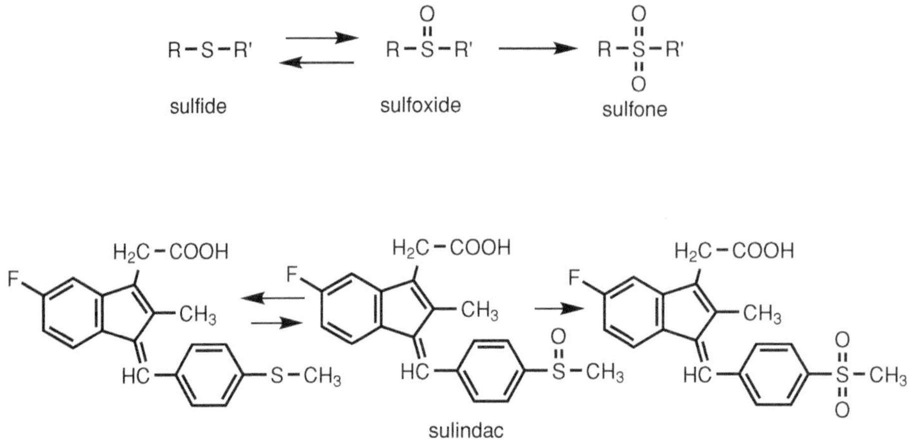

FIGURE 4.91 Oxidation of sulfides leads to sulfoxides followed by sulfones as shown with sulindac as an example.

between the drug and biological molecules as shown in Figure 4.90 (159). Sulfenic acids can be oxidized further to sulfinic acids and sulfonic acids (Fig. 4.90) (158) but this is not usually observed in vivo.

The first oxidation product of a sulfide is a sulfoxide and this can be further oxidized to a sulfone (Fig. 4.91). A good example is sulindac, which is a sulfoxide. It can be reduced to a sulfide or oxidized to a sulfone (Fig. 4.91). The sulfide is more active as a nonsteroidal anti-inflammatory agent than the parent drug but the sulfone is inactive (160).

REFERENCES

1. Schlichting I, Berendzen J, Chu K, et al. The catalytic pathway of cytochrome p450cam at atomic resolution. Science 2000;287(5458):1615–1622.
2. Gorsky LD, Koop DR, Coon MJ. On the stoichiometry of the oxidase and monooxygenase reactions catalyzed by liver microsomal cytochrome P-450. Products of oxygen reduction. J Biol Chem 1984;259(11):6812–6817.
3. Korzekwa KR, Jones JP. Predicting the cytochrome P450 mediated metabolism of xenobiotics. Pharmacogenetics 1993;3(1):1–18.
4. Groves JT, McClusky GA. Aliphatic hydroxylation by highly purified liver microsomal cytochrome P-450. Evidence for a carbon radical intermediate. Biochem Biophys Res Commun 1978;81(1):154–160.
5. Atkinson JK, Hollenberg PF, Ingold KU, et al. Cytochrome P450-catalyzed hydroxylation of hydrocarbons: kinetic deuterium isotope effects for the hydroxylation of an ultrafast radical clock. Biochemistry 1994;33(35):10630–10637.
6. Bowry VW, Ingold KU. A radical clock investigation of microsomal cytochrome-P-450 hydroxylation of hydrocarbons—rate of oxygen rebound. J Am Chem Soc 1991;113(15):5699–5707.
7. Newcomb M, Letadic MH, Putt DA, et al. An incredibly fast apparent oxygen rebound rate-constant for hydrocarbon hydroxylation by cytochrome-P-450 enzymes. J Am Chem Soc 1995;117(11):3312–3313.
8. Ogliaro F, Harris N, Cohen S, et al. A model "rebound" mechanism of hydroxylation by cytochrome P450: stepwise and effectively concerted pathways, and their reactivity patterns. J Am Chem Soc 2000;122(37):8977–8989.
9. Shimada T, Yamazaki H, Mimura M, et al. Interindividual variations in human liver cytochrome P-450 enzymes involved in the oxidation of drugs, carcinogens and toxic chemicals: studies

with liver microsomes of 30 Japanese and 30 Caucasians. J Pharmacol Exp Ther 1994;270(1): 414–423.

10. Miners JO, Mckinnon RA. CYP1A. In: Levy RH, Thummel KE, Trager WF, et al., eds. Metabolic Drug Interactions. Philadelphia, PA: Lippincott, Williams & Wilkins; 2000.

11. Pelkonen O, Rautio A, Raunio H, et al. CYP2A6: a human coumarin 7-hydroxylase. Toxicology 2000;144(1–3):139–147.

12. Nakajima M, Yamamoto T, Nunoya K, et al. Characterization of CYP2A6 involved in 3'-hydroxylation of cotinine in human liver microsomes. J Pharmacol Exp Ther 1996;277(2):1010–1015.

13. Tyndale RF, Sellers EM. Genetic variation in CYP2A6-mediated nicotine metabolism alters smoking behavior. Ther Drug Monit 2002;24(1):163–171.

14. Le Gal A, Dreano Y, Lucas D, et al. Diversity of selective environmental substrates for human cytochrome P450 2A6: alkoxyethers, nicotine, coumarin, N-nitrosodiethylamine, and N-nitrosobenzylmethylamine. Toxicol Lett 2003;144(1):77–91.

15. Chang TK, Weber GF, Crespi CL, et al. Differential activation of cyclophosphamide and ifosphamide by cytochromes P-450 2B and 3A in human liver microsomes. Cancer Res 1993;53(23):5629–5637.

16. Oda Y, Hamaoka N, Hiroi T, et al. Involvement of human liver cytochrome P4502B6 in the metabolism of propofol. Br J Clin Pharmacol 2001;51(3):281–285.

17. Hesse LM, Venkatakrishnan K, Court MH, et al. CYP2B6 mediates the in vitro hydroxylation of bupropion: potential drug interactions with other antidepressants. Drug Metab Dispos 2000;28(10):1176–1183.

18. Yuan R, Madani S, Wei XX, et al. Evaluation of cytochrome P450 probe substrates commonly used by the pharmaceutical industry to study in vitro drug interactions. Drug Metab Dispos 2002;30(12):1311–1319.

19. Black DJ, Kunze KL, Wienkers LC, et al. Warfarin-fluconazole. II.A metabolically based drug interaction: in vivo studies. Drug Metab Dispos 1996;24(4):422–428.

20. Rettie AE, Korzekwa KR, Kunze KL, et al. Hydroxylation of warfarin by human cDNA-expressed cytochrome P-450: a role for P-4502C9 in the etiology of (S)-warfarin-drug interactions. Chem Res Toxicol 1992;5(1):54–59.

21. Kunze KL, Eddy AC, Gibaldi M, et al. Metabolic enantiomeric interactions: the inhibition of human (S)-warfarin-7-hydroxylase by (R)-warfarin. Chirality 1991;3(1):24–29.

22. Veronese ME, Mackenzie PI, Doecke CJ, et al. Tolbutamide and phenytoin hydroxylations by cDNA-expressed human liver cytochrome P4502C9. Biochem Biophys Res Commun 1991;175(3):1112–1118.

23. Tracy TS, Rosenbluth BW, Wrighton SA, et al. Role of cytochrome P450 2C9 and an allelic variant in the 4'-hydroxylation of (R)- and (S)-flurbiprofen. Biochem Pharmacol 1995;49(9):1269–1275.

24. Leemann T, Transon C, Dayer P. Cytochrome P450TB (CYP2C): a major monooxygenase catalyzing diclofenac 4'-hydroxylation in human liver. Life Sci 1993;52(1):29–34.

25. Wester MR, Yano JK, Schoch GA, et al. The structure of human cytochrome P450 2C9 complexed with flurbiprofen at 2.0-A resolution. J Biol Chem 2004;279(34):35630–35637.

26. Rettie AE, Koop DR, Haining RL. CYP2C. In: Levy RH, Thummel KE, Trager WF, et al., eds. Metabolic Drug Interactions. Philadelphia, PA: Lippincott, Williams & Wilkins; 2000.

27. Haining RL, Hunter AP, Veronese ME, et al. Allelic variants of human cytochrome P450 2C9: baculovirus-mediated expression, purification, structural characterization, substrate stereoselectivity, and prochiral selectivity of the wild-type and I359L mutant forms. Arch Biochem Biophys 1996;333(2):447–458.

28. Steward DJ, Haining RL, Henne KR, et al. Genetic association between sensitivity to warfarin and expression of CYP2C9*3. Pharmacogenetics 1997;7(5):361–367.

29. Zanger UM, Eichelbaum M. CYP2D6. In: Levy RH, Thummel KE, Trager WF, et al., eds. Metabolic Drug Interactions. Philadelphia, PA: Lippincott, Williams & Wilkins; 2000.

30. Koymans LM, Vermeulen NP, Baarslag A, et al. A preliminary 3D model for cytochrome P450 2D6 constructed by homology model building. J Comput Aided Mol Des 1993;7(3):281–289.

31. Paine MJ, McLaughlin LA, Flanagan JU, et al. Residues glutamate 216 and aspartate 301 are key determinants of substrate specificity and product regioselectivity in cytochrome P450 2D6. J Biol Chem 2003;278(6):4021–4027.

32. van Waterschoot RA, Keizers PH, de Graaf C, et al. Topological role of cytochrome P450 2D6 active site residues. Arch Biochem Biophys 2006;447(1):53–58.

33. Rowland P, Blaney FE, Smyth MG, et al. Crystal structure of human cytochrome P450 2D6. J Biol Chem 2006;281(11):7614–7622.

34. Meyer UA, Zanger UM. Molecular mechanisms of genetic polymorphisms of drug metabolism. Annu Rev Pharmacol Toxicol 1997;37:269–296.

35. Raucy J, Carpenter SP. CYP2E1. In: Levy RH, Thummel KE, Trager WF, et al., eds. Metabolic Drug Interactions. Philadelphia, PA: Lippincott, Williams & Wilkins; 2000.

36. Koop DR. Oxidative and reductive metabolism by cytochrome P450 2E1. Faseb J 1992;6(2):724–730.

37. Mackman R, Guo Z, Guengerich FP, et al. Active site topology of human cytochrome P450 2E1. Chem Res Toxicol 1996;9(1):223–226.

38. Tan Y, White SP, Paranawithana SR, et al. A hypothetical model for the active site of human cytochrome P4502E1. Xenobiotica 1997;27(3):287–299.

39. Wrighton SA, Thummel KE. CYP3A. In: Levy RH, Thummel KE, Trager WF, et al., eds. Metabolic Drug Interactions. Philadelphia, PA: Lippincott, Williams & Wilkins; 2000.

40. Huang Z, Roy P, Waxman DJ. Role of human liver microsomal CYP3A4 and CYP2B6 in catalyzing N-dechloroethylation of cyclophosphamide and ifosfamide. Biochem Pharmacol 2000;59(8):961–972.

41. Marnett LJ, Kennedy TA. Comparison of the peroxidase activity of hemoproteins and cytochrome P450. In: Ortiz de Montellano PR, ed. Cytochrome P450 (2nd ed). New York, NY: Plenum; 1995:49–80.

42. Dunford HB. Horseradish peroxidase: structure and kinetic properties. In: Everse J, Everse KE, Grisham MB, eds. Peroxidases in Chemistry and Biology. Boca Raton, FL: CRC Press; 1991: 1–24.

43. Marnett LJ, Maddipati KR. Prostaglandin H synthase. In: Everse J, Everse KE, Grisham MB, eds. Peroxidases in Chemistry and Biology. Boca Raton, FL: CRC Press; 1991:293–334.

44. Zenser TV, Lakshmi VM, Hsu FF, et al. Metabolism of N-acetylbenzidine and initiation of bladder cancer. Mutat Res 2002;506–507:29–40.

45. Parman T, Chen G, Wells PG. Free radical intermediates of phenytoin and related teratogens. Prostaglandin H synthase-catalyzed bioactivation, electron paramagnetic resonance spectrometry, and photochemical product analysis. J Biol Chem 1998;273(39):25079–25088.

46. Parman T, Wells PG. Embryonic prostaglandin H synthase-2 (PHS-2) expression and benzo[a]pyrene teratogenicity in PHS-2 knockout mice. FASEB J 2002;16(9):1001–1009.

47. Marnett LJ. Prostaglandin synthase-mediated metabolism of carcinogens and a potential role for peroxyl radicals as reactive intermediates. Environ Health Perspect 1990;88:5–12.

48. Klebanoff SJ. Myeloperoxidase: occurence and biological function. In: Everse J, Everse KE, Grisham MB, eds. Peroxidases in Chemistry and Biology. Boca Raton, FL: CRC Press; 1991.

49. Budavari Se. The Merck Index (12th ed). Whitehouse Station, NJ: Merck Research Laboratories; 1996.

50. Liu ZC, Uetrecht JP. Clozapine is oxidized by activated human neutrophils to a reactive nitrenium ion that irreversibly binds to the cells. J Pharmacol Exp Ther 1995;275(3):1476–1483.

51. Maggs JL, Kitteringham NR, Breckenridge AM, et al. Autoxidative formation of a chemically reactive intermediate from amodiaquine, a myelotoxin and hepatotoxin in man. Biochem Pharmacol 1987;36(13):2061–2062.

52. Uetrecht JP, Ma HM, MacKnight E, et al. Oxidation of aminopyrine by hypochlorite to a reactive dication: possible implications for aminopyrine-induced agranulocytosis. Chem Res Toxicol 1995;8(2):226–233.

53. Uetrecht JP, Zahid N, Whitfield D. Metabolism of vesnarinone by activated neutrophils; implications for vesnarinone-induced agranulocytosis. J Pharmacol Exp Ther 1994;270(3):865–872.

54. Eling TE, Mason RP, Sivarajah K. The formation of aminopyrine cation radical by peroxidase activity of prostaglandin H synthase and subsequent reactions of the radical. J Biol Chem 1985;260(3):1601–1607.

55. Uetrecht J. Current trends in drug-induced autoimmunity. Autoimmun Rev 2005;4(5):309–314.

56. Gardner I, Leeder JS, Chin T, et al. A comparison of the covalent binding of clozapine and olanzapine to human neutrophils in vitro and in vivo. Mol Pharmacol 1998;53(6):999–1008.

57. Gorlewska-Roberts KM, Teitel CH, Lay JO, Jr., et al. Lactoperoxidase-catalyzed activation of carcinogenic aromatic and heterocyclic amines. Chem Res Toxicol 2004;17(12):1659–1966.

58. Josephy PD. The role of peroxidase-catalyzed activation of aromatic amines in breast cancer. Mutagenesis 1996;11(1):3–7.

59. Henderson WR, Jr. Eosinophil peroxidase: occurrence and biological function. In: Everse J, Everse KE, Grisham MB, eds. Peroxidases in Chemistry and Biology. Boca Raton, FL: CRC Press; 1991.

60. Taurog A, Dorris ML, Doerge DR. Minocycline and the thyroid: antithyroid effects of the drug, and the role of thyroid peroxidase in minocycline-induced black pigmentation of the gland. Thyroid 1996;6(3):211–219.

61. Cashman JR, Zhang J. Interindividual differences of human flavin-containing monooxygenase 3: genetic polymorphisms and functional variation. Drug Metab Dispos 2002;30(10):1043–1052.

62. Yeung CK, Lang DH, Thummel KE, et al. Immunoquantitation of FMO1 in human liver, kidney, and intestine. Drug Metab Dispos 2000;28(9):1107–1111.

63. Ziegler DM. Recent studies on the structure and function of multisubstrate flavin-containing monooxygenases. Annu Rev Pharmacol Toxicol 1993;33:179–199.

64. Beaty NB, Ballou DP. The reductive half-reaction of liver microsomal FAD-containing monooxygenase. J Biol Chem 1981;256(9):4611–4618.

65. Beaty NB, Ballou DP. The oxidative half-reaction of liver microsomal FAD-containing monooxygenase. J Biol Chem 1981;256(9):4619–4625.

66. Ziegler DM. An overview of the mechanism, substrate specificities, and structure of FMOs. Drug Metab Rev 2002;34(3):503–511.

67. He M, Rettie AE, Neal J, et al. Metabolism of sulfinpyrazone sulfide and sulfinpyrazone by human liver microsomes and cDNA-expressed cytochrome P450s. Drug Metab Dispos 2001;29(5): 701–711.

68. Lang DH, Yeung CK, Peter RM, et al. Isoform specificity of trimethylamine N-oxygenation by human flavin-containing monooxygenase (FMO) and P450 enzymes: selective catalysis by FMO3. Biochem Pharmacol 1998;56(8):1005–1012.

69. Reid JM, Walker DL, Miller JK, et al. The metabolism of pyrazoloacridine (NSC 366140) by cytochromes p450 and flavin monooxygenase in human liver microsomes. Clin Cancer Res 2004;10(4):1471–1480.

70. Cashman JR, Xiong YN, Xu L, et al. N-oxygenation of amphetamine and methamphetamine by the human flavin-containing monooxygenase (form 3): role in bioactivation and detoxication. J Pharmacol Exp Ther 1999;288(3):1251–1260.

71. Hamman MA, Haehner-Daniels BD, Wrighton SA, et al. Stereoselective sulfoxidation of sulindac sulfide by flavin-containing monooxygenases. Comparison of human liver and kidney microsomes and mammalian enzymes. Biochem Pharmacol 2000;60(1):7–17.

72. Duester G, Farres J, Felder MR, et al. Recommended nomenclature for the vertebrate alcohol dehydrogenase gene family. Biochem Pharmacol 1999;58(3):389–395.

73. Davani B, Khan A, Hult M, et al. Type 1 11beta -hydroxysteroid dehydrogenase mediates glucocorticoid activation and insulin release in pancreatic islets. J Biol Chem 2000;275(45):34841–34844.

74. Bosron WF, Li TK. Alcohol Dehydrogenase. In: Jakoby WB, ed. Enzymatic Basis of Detoxication. New York, NY: Academic Press; 1980:231–248.

75. Vasiliou V, Pappa A, Estey T. Role of human aldehyde dehydrogenases in endobiotic and xenobiotic metabolism. Drug Metab Rev 2004;36(2):279–299.

76. Weiner H. Aldehyde oxidizing enzymes. In: Jakoby WB, ed. Enzymatic Basis of Detoxication. New York, NY: Academic Press; 1980:261–280.

77. Vallari RC, Pietruszko R. Human aldehyde dehydrogenase: mechanism of inhibition of disulfiram. Science 1982;216(4546):637–639.

78. Petersen EN. The pharmacology and toxicology of disulfiram and its metabolites. Acta Psychiatr Scand Suppl 1992;369:7–13.

79. Sladek NE. Human aldehyde dehydrogenases: potential pathological, pharmacological, and toxicological impact. J Biochem Mol Toxicol 2003;17(1):7–23.

80. Duester G. Genetic dissection of retinoid dehydrogenases. Chem Biol Interact 2001;130–132 (1–3):469–480.

81. Bach AW, Lan NC, Johnson DL, et al. cDNA cloning of human liver monoamine oxidase A and B: molecular basis of differences in enzymatic properties. Proc Natl Acad Sci USA 1988;85(13):4934–4938.

82. Fowler CJ, Mantle TJ, Tipton KF. The nature of the inhibition of rat liver monoamine oxidase types A and B by the acetylenic inhibitors clorgyline, l-deprenyl and pargyline. Biochem Pharmacol 1982;31(22):3555–3561.

83. De Colibus L, Li M, Binda C, et al. Three-dimensional structure of human monoamine oxidase A (MAO A): relation to the structures of rat MAO A and human MAO B. Proc Natl Acad Sci USA 2005;102(36):12684–12689.

84. Binda C, Newton-Vinson P, Hubalek F, et al. Structure of human monoamine oxidase B, a drug target for the treatment of neurological disorders. Nat Struct Biol 2002;9(1):22–26.

85. Ottoboni S, Caldera P, Trevor A, et al. Deuterium isotope effect measurements on the interactions of the neurotoxin 1-methyl-4-phenyl-1,2,3,6-tetrahydropyridine with monoamine oxidase B. J Biol Chem 1989;264(23):13684–13688.

86. Edmondson DE, Mattevi A, Binda C, et al. Structure and mechanism of monoamine oxidase. Curr Med Chem 2004;11(15):1983–1993.

87. Castagnoli N, Jr., Trevor A, Singer TP, et al. Metabolic studies on the nigrostriatal toxin 1-methyl-4-phenyl-1,2,3,6-tetrahydropyridines. In: Sandler M, Dahlstrom A, eds. Progress in Catecholamine Research, Part B: Central Aspects. New York, NY: Alan R. Liss; 1988:93–100.

88. Mabic S, Castagnoli N, Jr. Assessment of structural requirements for the monoamine oxidase-B-catalyzed oxidation of 1,4-disubstituted-1,2,3,6-tetrahydropyridine derivatives related to the neurotoxin 1-methyl-4-phenyl-1,2,3,6-tetrahydropyridine. J Med Chem 1996;39(19):3694–3700.

89. Harrison R. Structure and function of xanthine oxidoreductase: where are we now? Free Radic Biol Med 2002;33(6):774–797.

90. Hille R. The mononuclear molybdenum enzymes. Chem Rev 1996;96(7):2757–2816.

91. Doonan CJ, Stockert A, Hille R, et al. Nature of the catalytically labile oxygen at the active site of xanthine oxidase. J Am Chem Soc 2005;127(12):4518–4522.

92. Bonini MG, Miyamoto S, Di Mascio P, et al. Production of the carbonate radical anion during xanthine oxidase turnover in the presence of bicarbonate. J Biol Chem 2004;279(50):51836–51843.

93. Huber R, Hof P, Duarte RO, et al. A structure-based catalytic mechanism for the xanthine oxidase family of molybdenum enzymes. Proc Natl Acad Sci USA 1996;93(17):8846–8851.

94. Panoutsopoulos GI, Beedham C. Kinetics and specificity of guinea pig liver aldehyde oxidase and bovine milk xanthine oxidase towards substituted benzaldehydes. Acta Biochim Pol 2004;51(3):649–663.

95. Beedham C. Molybdenum hydroxylases as drug-metabolizing enzymes. Drug Metab Rev 1985;16(1–2):119–156.

96. Rooseboom M, Commandeur JN, Vermeulen NP. Enzyme-catalyzed activation of anticancer prodrugs. Pharmacol Rev 2004;56(1):53–102.

97. Powell PK, Wolf I, Lasker JM. Identification of CYP4A11 as the major lauric acid omega-hydroxylase in human liver microsomes. Arch Biochem Biophys 1996;335(1):219–226.

98. Sawamura A, Kusunose E, Satouchi K, et al. Catalytic properties of rabbit kidney fatty acid omega-hydroxylase cytochrome P-450ka2 (CYP4A7). Biochim Biophys Acta 1993;1168(1):30–36.

99. Alexander JJ, Snyder A, Tonsgard JH. Omega-oxidation of monocarboxylic acids in rat brain. Neurochem Res 1998;23(2):227–233.

100. Kikuta Y, Kusunose E, Kusunose M. Characterization of human liver leukotriene B(4) omega-hydroxylase P450 (CYP4F2). J Biochem (Tokyo) 2000;127(6):1047–1052.

101. Powell PK, Wolf I, Jin R, et al. Metabolism of arachidonic acid to 20-hydroxy-5,8,11, 14-eicosatetraenoic acid by P450 enzymes in human liver: involvement of CYP4F2 and CYP4A11. J Pharmacol Exp Ther 1998;285(3):1327–1336.

102. Aitken AE, Roman LJ, Loughran PA, et al. Expressed CYP4A4 metabolism of prostaglandin E(1) and arachidonic acid. Arch Biochem Biophys 2001;393(2):329–338.

103. Payne AH. Hormonal regulation of cytochrome P450 enzymes, cholesterol side-chain cleavage and 17 alpha-hydroxylase/C17–20 lyase in Leydig cells. Biol Reprod 1990;42(3):399–404.

104. Hjelmeland LM, Aronow L, Trudell JR. Intramolecular determination of substituent effects in hydroxylations catalyzed by cytochrome P-450. Mol Pharmacol 1977;13(4):634–639.

105. Higgins L, Bennett GA, Shimoji M, et al. Evaluation of cytochrome P450 mechanism and kinetics using kinetic deuterium isotope effects. Biochemistry 1998;37(19):7039–7046.

106. White RE, Miller JP, Favreau LV, et al. Stereochemical dynamics of aliphatic hydroxylation by cytochrome-P-450. J Am Chem Soc 1986;108(19):6024–6031.

107. Korzekwa KR, Trager WF, Gillette JR. Theory for the observed isotope effects from enzymatic systems that form multiple products via branched reaction pathways: cytochrome P-450. Biochemistry 1989;28(23):9012–9018.

108. Higgins L, Korzekwa KR, Rao S, et al. An assessment of the reaction energetics for cytochrome P450-mediated reactions. Arch Biochem Biophys 2001;385(1):220–230.

109. Jones JP, Mysinger M, Korzekwa KR. Computational models for cytochrome P450: a predictive electronic model for aromatic oxidation and hydrogen atom abstraction. Drug Metab Dispos 2002;30(1):7–12.

110. Licht HJ, Coscia CJ. Cytochrome P-450LM2 mediated hydroxylation of monoterpene alcohols. Biochemistry 1978;17(26):5638–5646.

111. Shirane N, Sui Z, Peterson JA, et al. Cytochrome P450BM-3 (CYP102): regiospecificity of oxidation of omega-unsaturated fatty acids and mechanism-based inactivation. Biochemistry 1993;32(49):13732–13741.

112. Halpin RA, Ulm EH, Till AE, et al. Biotransformation of lovastatin.V. Species differences in in vivo metabolite profiles of mouse, rat, dog, and human. Drug Metab Dispos 1993;21(6):1003–1011.

113. Groves JT, Subramanian DV. Hydroxylation by cytochrome-P-450 and metalloporphyrin models—evidence for allylic rearrangement. J Am Chem Soc 1984;106(7):2177–2181.

114. Mcclanahan RH, Huitric AC, Pearson PG, et al. Evidence for a cytochrome-P-450 catalyzed allylic rearrangement with double-bond topomerization. J Am Chem Soc 1988;110(6):1979–1981.

115. Korzekwa KR, Jones JP, Gillette JR. Theoretical-studies on cytochrome-P-450 mediated hydroxylation—a predictive model for hydrogen-atom abstractions. J Am Chem Soc 1990;112(19):7042–7046.

116. Breck GD, Trager WF. Oxidative N-dealkylation: a mannich intermediate in the formation of a new metabolite of lidocaine in man. Science 1971;173(996):544–546.

117. Miwa GT, Walsh JS, Kedderis GL, et al. The use of intramolecular isotope effects to distinguish between deprotonation and hydrogen atom abstraction mechanisms in cytochrome P-450- and peroxidase-catalyzed N-demethylation reactions. J Biol Chem 1983;258(23):14445–14449.

118. Nelson SD, Breck GD, Trager WF. In vivo metabolite condensations. Formation of N1-ethyl-2-methyl-N3-(2,6-dimethylphenyl)-4-imidazolidinone from the reaction of a metabolite of alcohol with a metabolite of lidocaine. J Med Chem 1973;16(10):1106–1112.

119. Shen T, Hollenberg PF. The mechanism of stimulation of NADPH oxidation during the mechanism-based inactivation of cytochrome P450 2B1 by N-methylcarbazole: redox cycling and DNA scission. Chem Res Toxicol 1994;7(2):231–238.

120. Galliani G, Nali M, Rindone B, et al. The rate of N-demethylation of N, N-dimethylanilines and N-methylanilines by rat-liver microsomes is related to their first ionization potential, their lipophilicity and to a steric bulk factor. Xenobiotica 1986;16(6):511–517.

121. Hall LR, Hanzlik RP. N-dealkylation of tertiary amides by cytochrome P-450. Xenobiotica 1991;21(9):1127–1138.

122. Clement B, Zimmermann M. Hepatic microsomal N-demethylation of N-methylbenzamidine. N-d ealkylation vs N-oxygenation of amidines. Biochem Pharmacol 1987;36(19):3127–3133.

123. Keefer LK, Anjo T, Wade D, et al. Concurrent generation of methylamine and nitrite during denitrosation of N-nitrosodimethylamine by rat liver microsomes. Cancer Res 1987;47(2):447–452.

124. Satoh H, Fukuda Y, Anderson DK, et al. Immunological studies on the mechanism of halothane-induced hepatotoxicity: immunohistochemical evidence of trifluoroacetylated hepatocytes. J Pharmacol Exp Ther 1985;233(3):857–862.

125. Spracklin DK, Hankins DC, Fisher JM, et al. Cytochrome P450 2E1 is the principal catalyst of human oxidative halothane metabolism in vitro. J Pharmacol Exp Ther 1997;281(1):400–411.

126. Christ DD, Kenna JG, Kammerer W, et al. Enflurane metabolism produces covalently bound liver adducts recognized by antibodies from patients with halothane hepatitis. Anesthesiology 1988;69(6):833–838.

127. Martin HJ, Breyer-Pfaff U, Wsol V, et al. Purification and characterization of akr1b10 from human liver: role in carbonyl reduction of xenobiotics. Drug Metab Dispos 2006;34(3):464–470.

128. Harris JW, Pohl LR, Martin JL, et al. Tissue acylation by the chlorofluorocarbon substitute 2,2-dichloro-1,1,1-trifluoroethane. Proc Natl Acad Sci USA 1991;88(4):1407–1410.

129. Yin H, Jones JP, Anders MW. Metabolism of 1-fluoro-1,1,2-trichloroethane, 1,2-dichloro-1,1-difluoroethane, and 1,1,1-trifluoro-2-chloroethane. Chem Res Toxicol 1995;8(2):262–268.

130. Yin H, Anders MW, Korzekwa KR, et al. Designing safer chemicals: predicting the rates of metabolism of halogenated alkanes. Proc Natl Acad Sci USA 1995;92(24):11076–11080.

131. Rietjens IM, den Besten C, Hanzlik RP, et al. Cytochrome P450-catalyzed oxidations of halobenzene. Chem Res Toxicol 1997;10(6):629–635.

132. Guengerich FP, Peterson LA, Bocker RH. Cytochrome P-450-catalyzed hydroxylation and carboxylic acid ester cleavage of Hantzsch pyridine esters. J Biol Chem 1988;263(17):8176–8183.

133. Peng HM, Raner GM, Vaz AD, et al. Oxidative cleavage of esters and amides to carbonyl products by cytochrome P450. Arch Biochem Biophys 1995;318(2):333–339.

134. Grogan J, DeVito SC, Pearlman RS, et al. Modeling cyanide release from nitriles: prediction of cytochrome P450 mediated acute nitrile toxicity. Chem Res Toxicol 1992;5(4):548–552.

135. Rettie AE, Boberg M, Rettenmeier AW, et al. Cytochrome P-450-catalyzed desaturation of valproic acid in vitro. Species differences, induction effects, and mechanistic studies. J Biol Chem 1988;263(27):13733–13738.

136. Korzekwa KR, Trager WF, Nagata K, et al. Isotope effect studies on the mechanism of the cytochrome P-450IIA1-catalyzed formation of delta 6-testosterone from testosterone. Drug Metab Dispos 1990;18(6):974–979.

137. Obach RS. Mechanism of cytochrome P4503A4- and 2D6-catalyzed dehydrogenation of ezlopitant as probed with isotope effects using five deuterated analogs. Drug Metab Dispos 2001;29(12):1599–1607.

138. Ortiz de Montellano PR, Reich NO. Inhibition of cytochrome P-450 enzymes. In: Ortiz de Montellano PR, ed. Cytochrome P-450 (1st ed). New York, NY: Plenum; 1986.

139. Shaik S, Kumar D, de Visser SP, et al. Theoretical perspective on the structure and mechanism of cytochrome P450 enzymes. Chem Rev 2005;105(6):2279–2328.

140. Ortiz de Montellano PR, Voss JJ. Substrate oxidation by cytochrome P450. In: Ortiz de Montellano PR, ed. Cytochrome P-450, Structure, Mechanism, and Biochemistry (3rd ed). New York, NY: Kluwer Academic/Plenum; 2005:198–200.

141. Ortiz de Montellano PR, Komives EA. Branchpoint for heme alkylation and metabolite formation in the oxidation of arylacetylenes by cytochrome P-450. J Biol Chem 1985;260(6):3330–3336.

142. Kent UM, Mills DE, Rajnarayanan RV, et al. Effect of 17-alpha-ethynylestradiol on activities of cytochrome P450 2B (P450 2B) enzymes: characterization of inactivation of P450s 2B1 and 2B6 and identification of metabolites. J Pharmacol Exp Ther 2002;300(2):549–558.

143. Jerina DM, Daly JW. Arene oxides: a new aspect of drug metabolism. Science 1974;185(151):573–582.

144. Korzekwa K, Trager W, Gouterman M, et al. Cytochrome-P450 mediated aromatic oxidation—a theoretical-study. J Am Chem Soc 1985;107(14):4273–4279.

145. Hanzlik RP, Hogberg K, Judson CM. Microsomal hydroxylation of specifically deuterated monosubstituted benzenes. Evidence for direct aromatic hydroxylation. Biochemistry 1984;23(13):3048–3055.

146. Bush ED, Trager WF. Substrate probes for the mechanism of aromatic hydroxylation catalyzed by cytochrome P-450: selectively deuterated analogues of warfarin. J Med Chem 1985;28(8):992–996.

147. Darbyshire JF, Iyer KR, Grogan J, et al. Substrate probe for the mechanism of aromatic hydroxylation catalyzed by cytochrome P450. Drug Metab Dispos 1996;24(9):1038–1045.

148. de Visser SP, Shaik S. A proton-shuttle mechanism mediated by the porphyrin in benzene hydroxylation by cytochrome P450 enzymes. J Am Chem Soc 2003;125(24):7413–7424.

149. Ohe T, Mashino T, Hirobe M. Substituent elimination from p-substituted phenols by cytochrome P450. Ipso-substitution by the oxygen atom of the active species. Drug Metab Dispos 1997;25(1):116–122.

150. Vaz ADN, Roberts ES, Coon MJ. Olefin Formation in the oxidative deformylation of aldehydes by cytochrome-P-450—mechanistic implications for catalysis by oxygen-derived peroxide. J Am Chem Soc 1991;113(15):5886–5887.

151. Roberts ES, Vaz AD, Coon MJ. Catalysis by cytochrome P-450 of an oxidative reaction in xenobiotic aldehyde metabolism: deformylation with olefin formation. Proc Natl Acad Sci USA 1991;88(20):8963–8966.

152. Zhang Z, Li Y, Stearns RA, et al. Cytochrome P450 3A4-mediated oxidative conversion of a cyano to an amide group in the metabolism of pinacidil. Biochemistry 2002;41(8):2712–2718.

153. Mangan FR, Flack JD, Jackson D. Preclinical overview of nabumetone. Pharmacology, bioavailability, metabolism, and toxicology. Am J Med 1987;83(4B):6–10.

154. Tweedie DJ, Erikson JM, Prough RA. Metabolism of hydrazine anti-cancer agents. Pharmacol Ther 1987;34(1):111–127.

155. Hofstra A, Uetrecht JP. Metabolism of hydralazine to a reactive intermediate by the oxidizing system of activated leukocytes. Chemico-Biol Interact 1993;89:183–196.

156. Hofstra AH, Li-Muller SM, Uetrecht JP. Metabolism of isoniazid by activated leukocytes. Possible role in drug-induced lupus. Drug Metab Dispos 1992;20(2):205–210.

157. Seto Y, Guengerich FP. Partitioning between N-dealkylation and N-oxygenation in the oxidation of N, N-dialkylarylamines catalyzed by cytochrome P450 2B1. J Biol Chem 1993;268(14):9986–9997.

158. Testa B, Jenner P. Drug Metabolism: Chemical and Biochemical Aspects. New York, NY: Dekker; 1976.

159. Migdalof BH, Antonaccio MJ, McKinstry DN, et al. Captopril: pharmacology, metabolism and disposition. Drug Metab Rev 1984;15(4):841–869.

160. Mattila J, Mantyla R, Vuorela A, et al. Pharmacokinetics of graded oral doses of sulindac in man. Arzneimittelforschung 1984;34(2):226–229.

5
Reductive Pathways

Reduction is the reverse of oxidation and therefore it can involve loss of an oxygen atom or the addition of two hydrogen atoms. Many of the same enzymes that are involved in oxidation can also mediate reductions. For example, some drugs, especially those that are very electron deficient because of nitro groups, etc., can be reduced by cytochromes P450. Some other enzymes, such as alcohol dehydrogenase (ALD) are readily reversible and the same enzyme can also catalyze reduction. Much reduction occurs in anaerobic bacteria in the gut because, being anaerobic, much of the metabolism of these organisms involves reductive pathways.

MAJOR REDUCING ENZYMES

Aldo/Keto Reductases

The aldo/keto reductases (AKR) constitute a superfamily of soluble oxidoreductases. They occur in most living organisms and utilize $NADP^+(H)$ as cofactor to reduce aldehydes and ketones to primary and secondary alcohols, respectively. The reactions are reversible, but unlike the ALDs, the AKRs generally catalyze reductive rather than oxidative reactions. To date, 14 families of AKRs have been identified that together contain over 100 different proteins (1). Out of these, at least eight individual proteins are human enzymes and members of one of two families, AKR1 and AKR7 (AKR1B1, AKR1B10, AKR1C1, AKR1C2, AKR1C3, AKR1C4, AKR7A2, and AKR7A3) (1). In general, the AKRs have broad substrate selectivity and tend to operate on both endobiotics (e.g., carbohydrates, ketosteroids, retinal) and xenobiotics [e.g., the antiemetic agent dolasetron, the antitumor drug daunorubicin, and the tobacco-specific carcinogen 4-methylnitrosoamine-1-(3-pyridyl)-1-butanone (NNK)] (Fig. 5.1) (2). At least four AKR isoforms (AKR1C1, 1C2, 1C4, and 1B10) have been isolated from human liver cytosol that contribute to the reduction of NNK to the less toxic 4-methylnitrosoamine-1-(3-pyridyl)-1-butanol, which is susceptible to conjugation and elimination.

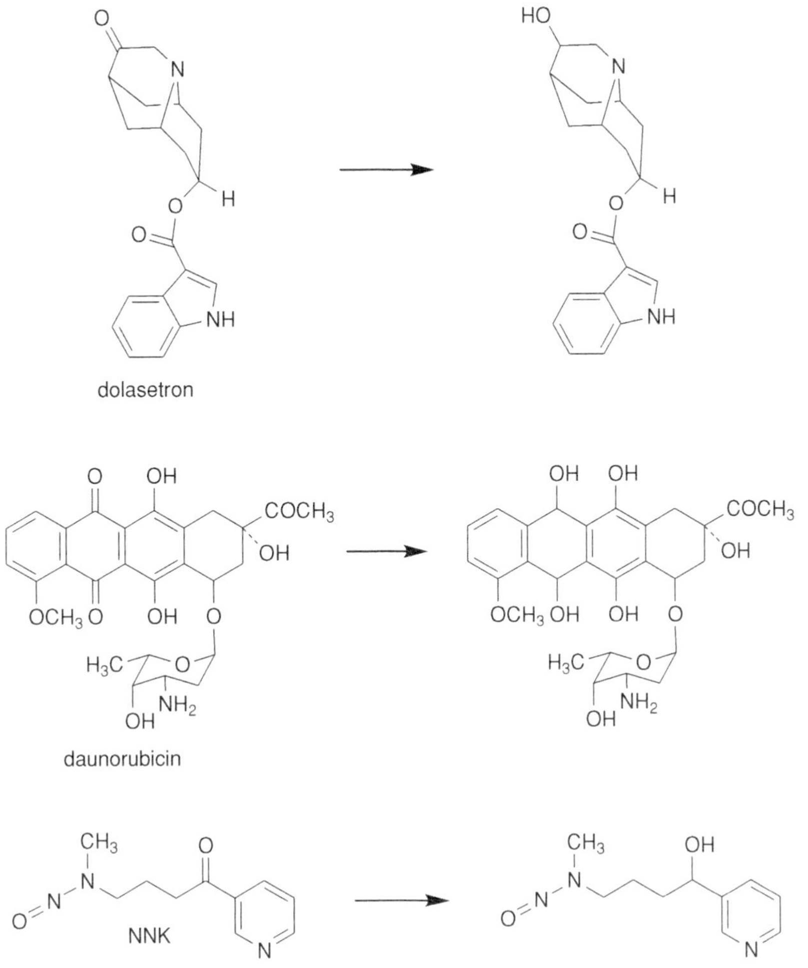

FIGURE 5.1　Reduction of specific substrates for aldo/keto reductases.

More recently, the reductions of the opioid receptor antagonist naltrexone and the antiemetic agent dolasetron were investigated with the human AKRs—AKR1C1, 1C2, and 1C4 (3). All three isoforms were able to reduce both substrates. However, AKR1C4 was 1000 times more efficient (Vmax/Km) than AKR1C1 in reducing naltrexone, while AKR1C2 was intermediate in efficiency. AKR1C1 and AKR1C4 were the most efficient in reducing dolasetron. The authors conclude that the efficient reduction of naltrexone by AKR1C4 is probably responsible for the high ratios of 6-β-naltrexol/naltrexone seen in the human.

AKR7A2 and perhaps the highly homologous (88%) AKR7A3 have been identified as the succinic semialdehyde reductase responsible for the biosynthesis of the neuromodulator γ-hydroxybutyric acid from succinic semialdehyde (4). Both proteins may also play a significant role in reducing damage to the brain from the aldehydes produced by stress-induced lipid peroxidation that underlies many neurodegenerative disorders.

The main human AKRs have been cloned, expressed, and their substrate reactivity profiles and tissue and organ distribution determined.

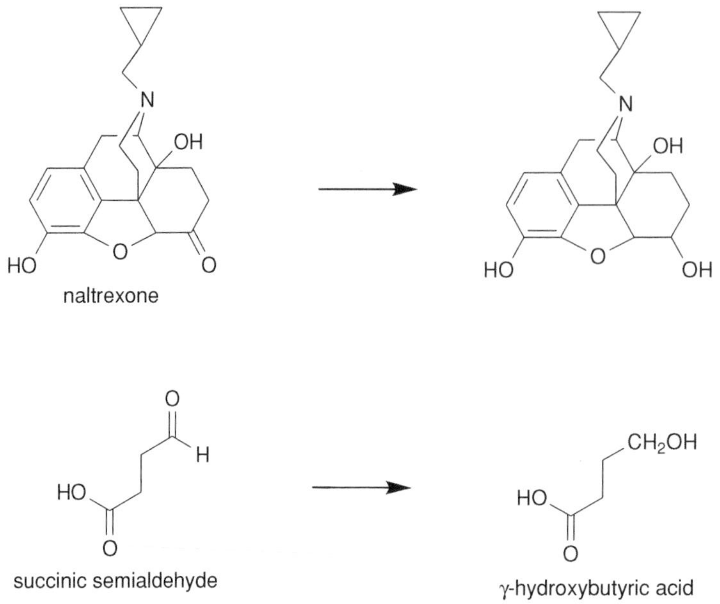

naltrexone

succinic semialdehyde

γ-hydroxybutyric acid

FIGURE 5.2 Additional substrates for aldo/keto reductases.

Carbonyl Reductases

Carbonyl reductases (CBRs) are a NADPH-dependent subset of enzymes of a super family of oxidoreductases that, in turn, are a subset of the more extensive short-chain dehydrogenase/reductases (1). Like the AKRs, they are found in cytosol and have broad substrate selectivity that encompasses both endogenous and xenobiotic carbonyl compounds including prostaglandins, steroids, quinones, in addition to a wide array of aromatic and aliphatic aldehydes and ketones (1,5). To date, two human isoforms, CBR1 and CBR3, have been identified and characterized while a third, CBR4, may also be present based on genomic analysis but is as yet uncharacterized (1).

A few typical examples of CBR-catalyzed reduction of xenobiotics include the reduction of the antipsychotic haloperidol (6), the P450 inhibitor metyrapone (6), the oral hypoglycemic acetohexamide (6), and the anticoagulant warfarin (7).

Cytochrome P450

The catalytic activity of cytochrome P450 is not restricted to oxidation. Under certain conditions, especially anaerobic conditions or with certain substrates, it can function as a reductase. For example, P450 can catalyze the reductive removal of halide from polyhalogenated alkanes such as hexachloroethane or halothane (8,9).

The mechanism is thought to involve two sequential transfers of an electron from reduced P450 to the halogenated hydrocarbon (Fig. 5.4). The first electron adds to a halide atom, which then eliminates as a halide anion. The second electron is transferred to the residual carbon radical generating a carbon anion. The carbon anion has a number of options. It can protonate to generate the monodehalogenated alkane, or it can undergo β-elimination of a second halide forming the didehalogenated alkene (8,9).

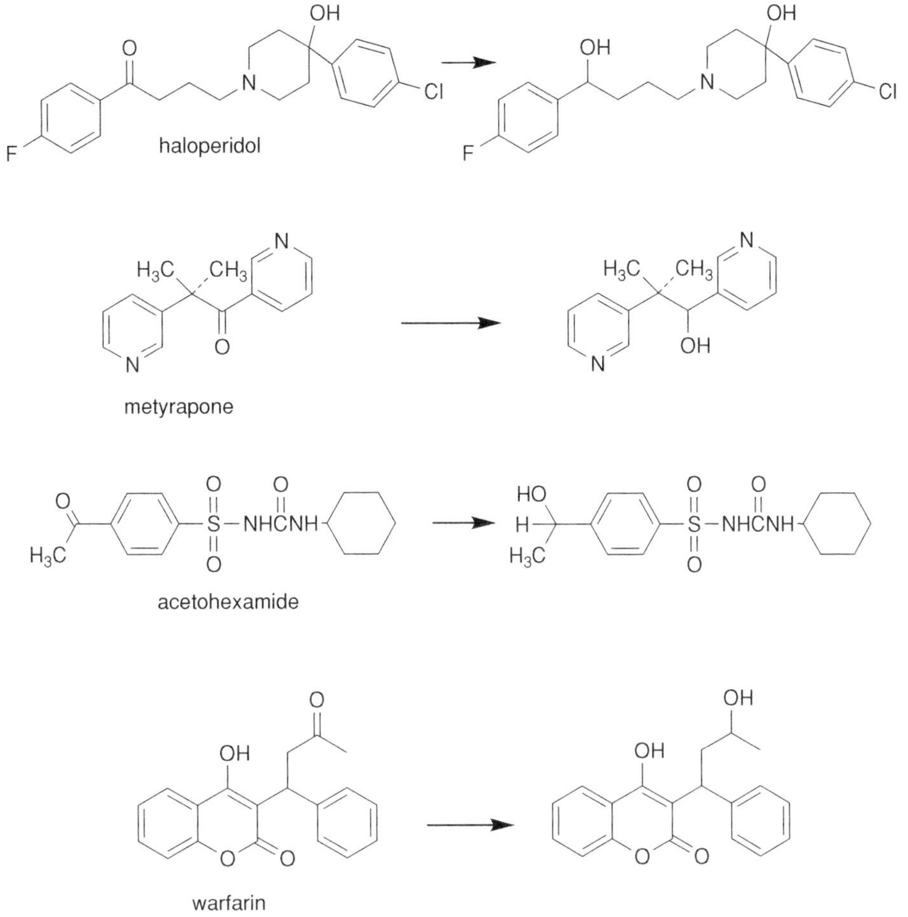

FIGURE 5.3 Reduction of substrates for carbonyl reductases.

Carbon tetrachloride is a solvent that is chemically inert, highly resistant to oxidation, but biologically toxic. Despite its chemical stability, P450 is able to convert carbon tetra-chloride to several reactive species. Reduced P450 transfers an electron to chloride leading to the elimination of a chloride anion and the generation of the reactive trichloromethyl radical (10). Trichloromethyl radical can undergo a second one-electron reduction to

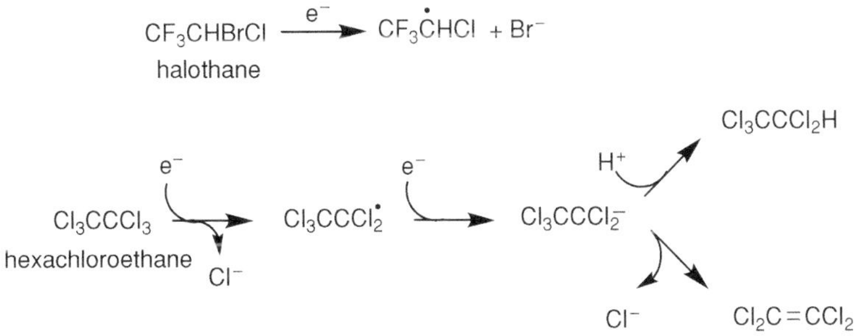

FIGURE 5.4 Examples of P450-catalyzed reductive dehalogenation.

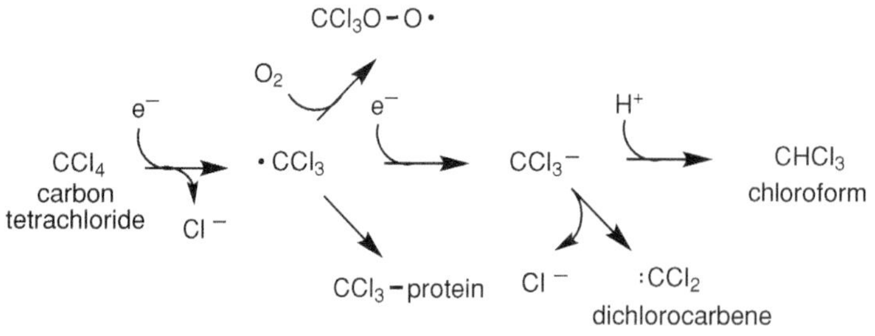

FIGURE 5.5 P450-mediated reduction of carbon tetrachloride and subsequent reactions of the trichloromethyl radical.

generate the trichloromethyl anion followed by protonation to yield chloroform. Alternatively, trichloromethyl radical can adduct to the protein or generate two additional reactive species, either by reacting with molecular oxygen to form trichloromethylperoxy radical or eliminating a second chloride anion to generate dichlorocarbene (Fig. 5.5).

Reductive dehalogenation is not the only reductive reaction catalyzed by cytochrome P450. Nitrogen-containing functional groups of various oxidation states can also be reduced back to the corresponding saturated nitrogen-containing functional group, e.g., amine, hydrazine, amidine, etc. Therefore, N-oxides, imines, hydroxylamines, nitroso groups, nitro groups, and azo dyes are all susceptible to reduction by P450, particularly under anaerobic conditions.

Xanthine Oxidase and Aldehyde Oxidase

Cytochrome P450-containing enzyme systems are not the only enzymes that are effective in reducing nitrogen-containing functional groups. In fact, they frequently are not the major source of reduction. The cytosolic molybdenum-containing enzymes, xanthine oxidase and particularly aldehyde oxidase , are often the major contributors. For example, they are major contributors in the reduction of aromatic nitro compounds (11,12) and perhaps N-oxides (13) to the corresponding amines.

NAD(P)H Quinone Oxidoreductase

Quinones are very common reactive metabolites of drugs and other xenobiotics. Although quinones appear structurally similar to ketones, their reduction is not usually mediated by keto reductases. The major enzyme responsible for this reduction is NAD(P)H quinone oxidoreductase, until recently known as DT-diaphorase and now termed NQO1 or QR1 (14). A second form of the enzyme, NQO2or QR2, was characterized almost 50 years ago, then forgotten until it reemerged in the early 1990s (15). The two enzymes are widely distributed throughout the body, but expression varies considerably with individual, organ, and physiological state. QR2 is a homodimer of 230 amino acids per monomer, while the monomer of QR1 contains an additional 44 amino acid residues. Both enzymes are flavin dependent but utilize different reducing cofactors to transfer electrons via hydride to flavin adenine dinucleotide (FAD) for further passage to the quinone substrate. While QR1 can utilize either NADH or NADPH in this regard, QR2 cannot. QR2 requires reduced *N*-ribosyl or *N*-alkyldihydronicotinamide as the source of reducing equivalents (Fig. 5.6).

What distinguishes QR1 and QR2 from other quinone-reducing enzymes, such as cytochrome P450, is that they are a direct two-electron reductant, i.e., quinone is reduced

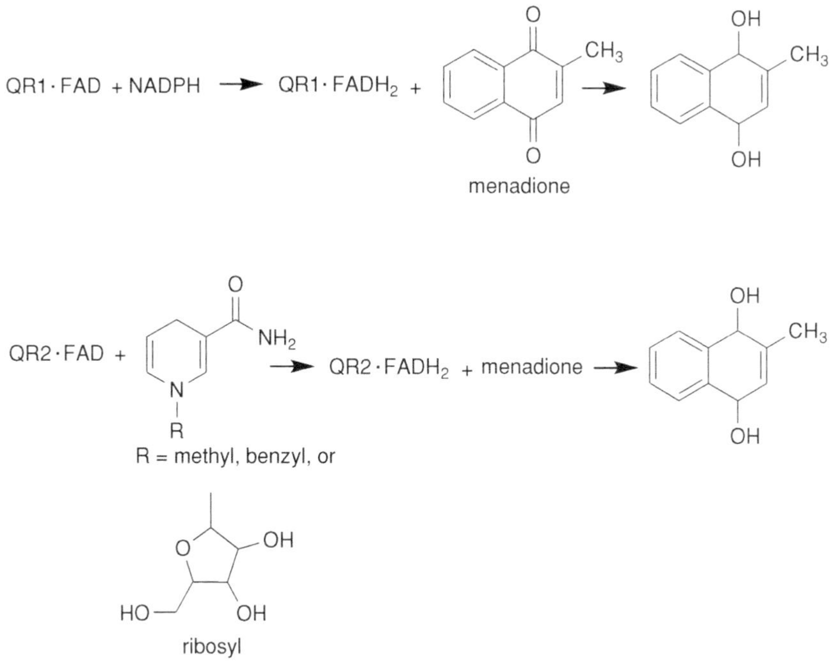

$$QR1 \cdot FAD + NADPH \longrightarrow QR1 \cdot FADH_2 +$$

menadione

$$QR2 \cdot FAD + \quad\quad\quad \longrightarrow QR2 \cdot FADH_2 + menadione \longrightarrow$$

R = methyl, benzyl, or

ribosyl

FIGURE 5.6 Mechanism for the reduction of menadione by QR1or QR2.

to hydroquinone in a single step. This has important ramifications in terms of toxicity. The two-electron reduction leads directly in a single step to the hydroquinone, a species that is inherently less toxic by virtue of being chemically less reactive. In contrast, P450 reduction of quinones takes place in two one-electron steps. This leads to the necessary formation of the chemically reactive semiquinone radical as an intermediate on the road to the final product, the hydroquinone. So clearly, out of the two enzymatic processes QR catalysis presents the less toxic liability. Consistent with its protective role against oxidative stress, NQO1 expression is controlled through the Nrf2/antioxidant response element similar to the control of glutathione transferases (16).

REDUCTIVE METABOLIC PATHWAYS

Reduction of Nitro, Nitroso, and Hydroxylamine Groups

Nitro groups can be reduced all the way to amines (Fig. 5.7). The first step requires an enzyme such as cytochrome P450 or anaerobic bacteria, but reduction of nitroso groups is so facile it is usually a simple chemical reduction mediated by biological reducing agents such as ascorbate or NADPH. Although the pathways are shown as two-electron oxidations and reductions, one-electron chemistry can also occur.

There are not many drugs that contain a nitro group and most are toxic. Many of the metabolites of nitro groups are the same as those of amines, and the reversibility of these

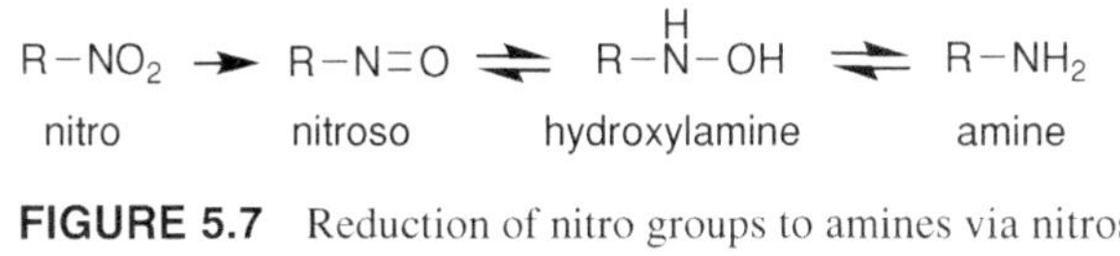

$$R-NO_2 \longrightarrow R-N{=}O \rightleftharpoons R-\overset{\text{H}}{N}-OH \rightleftharpoons R-NH_2$$

  nitro   nitroso   hydroxylamine   amine

FIGURE 5.7 Reduction of nitro groups to amines via nitroso and hydroxylamine intermediates.

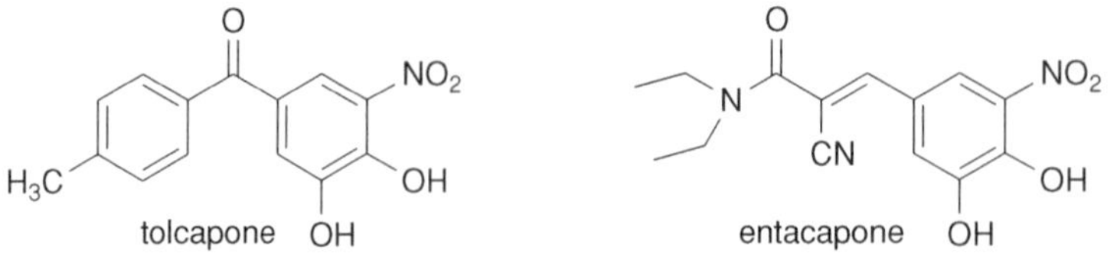

FIGURE 5.8 Structures of tolcapone and entacapone —only the nitro group of tolcapone is extensively reduced in vivo.

pathways provides the potential for redox cycling and the generation of reactive oxygen species as discussed in chapter 8. There appears to be a relationship between the toxicity of drugs containing a nitro group and the degree to which the nitro group is reduced, e.g., tolcapone is hepatotoxic and entacapone is not, and the nitro group of tolcapone is reduced to a greater degree than that of entacapone (Fig. 5.8) (17).

Reduction of Amine Oxides

Amine oxides are readily reduced back to tertiary amines (Fig. 5.9). There are few drugs that are amine oxides, but there are many drugs that are tertiary amines and amine oxides are common metabolites. The amine oxide is often pharmacologically inactive; however, because they are readily reduced back to tertiary amines, amine oxides can act as a "buffer" to the concentration of the tertiary amine.

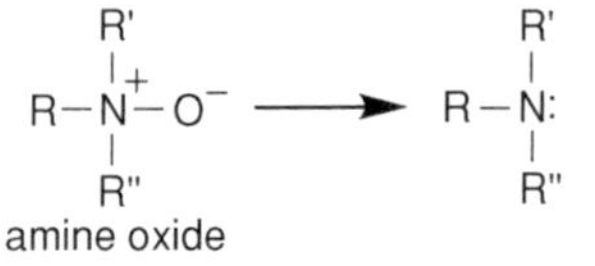

FIGURE 5.9 Reduction of tertiary amine oxides to tertiary amines.

Reduction of Azo Compounds

A four-electron reduction of the azo group leads to the cleavage of the molecule and the production of two amines (Fig. 5.10). There are few drugs that contain an azo bond but a good example is sulfasalazine, which is reductively cleaved to 4-aminosalicylic acid and sulfapyridine (18). This reduction is mediated by anaerobic bacteria in the intestine, and it leads to the formation of two agents that are pharmacologically active in the treatment of ulcerative colitis.

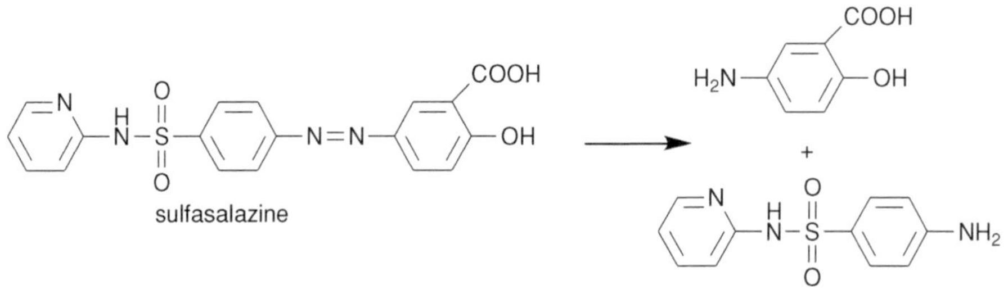

FIGURE 5.10 Reduction of azo-containing drugs leads to two amines.

Reduction of Aldehydes and Ketones

Aldehydes and ketones are readily reduced back to primary and secondary alcohols, respectively. In the case of ketones, although the reduction is reversible, ketoreductase utilizes NADPH, the concentration of which is higher than NADP+, and this drives the reaction toward the secondary alcohol. A good example is warfarin as shown in Figure 5.3 (19). However, aldehydes are further oxidized to carboxylic acids and *carboxylic acids are not reduced back to aldehydes* thus eliminating the aldehyde. Reductive metabolism of esters and amides also does not generally occur.

The reduction of aldehydes is not usually apparent because aldehydes are generally rapidly oxidized and oxidation to carboxylic acids is basically an irreversible process. Aldehydes with electron-withdrawing groups, however, such as trifluoroacetaldehyde, are more readily reduced since they are less readily oxidized and therefore this pathway is more evident.

Reduction of Quinones

Quinones are formed by the oxidation of hydroquinones and are readily reduced back to hydroquinones as shown in Figure 5.11. They are a chemically reactive functional group (see chap. 8) that does not normally occur in drug molecules. However, quinones are a relatively common environmental pollutant that arises from the burning of organics. In addition, humans are exposed through food intake, automobile exhaust, and cigarette smoke. A para quinone is shown in Figure 5.11, but the same is true of ortho quinones. As discussed earlier in this chapter, the major enzymes involved in this reduction are NQO1 and NQO2.

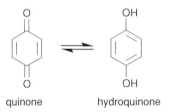

quinone   hydroquinone

FIGURE 5.11   Reversible reduction of quinones to hydroquinones.

Although reduction of quinones is usually a detoxication pathway, there are examples such as mitomycin C in which the hydroquinone is more toxic than the quinone as shown in Figure 5.12 and this may increase the susceptibility of cancers that express high levels of NQO. In this case, the reduction of the quinone leads to the loss of methanol, which is the first step in the activation of this anticancer agent (20).

Reduction of Sulfoxides

Sulfoxides are readily reduced to sulfides; however, analogous to the oxidation aldehydes, the oxidation of sulfoxides to sulfones is irreversible as illustrated by the drug sulindac in Figure 5.13.

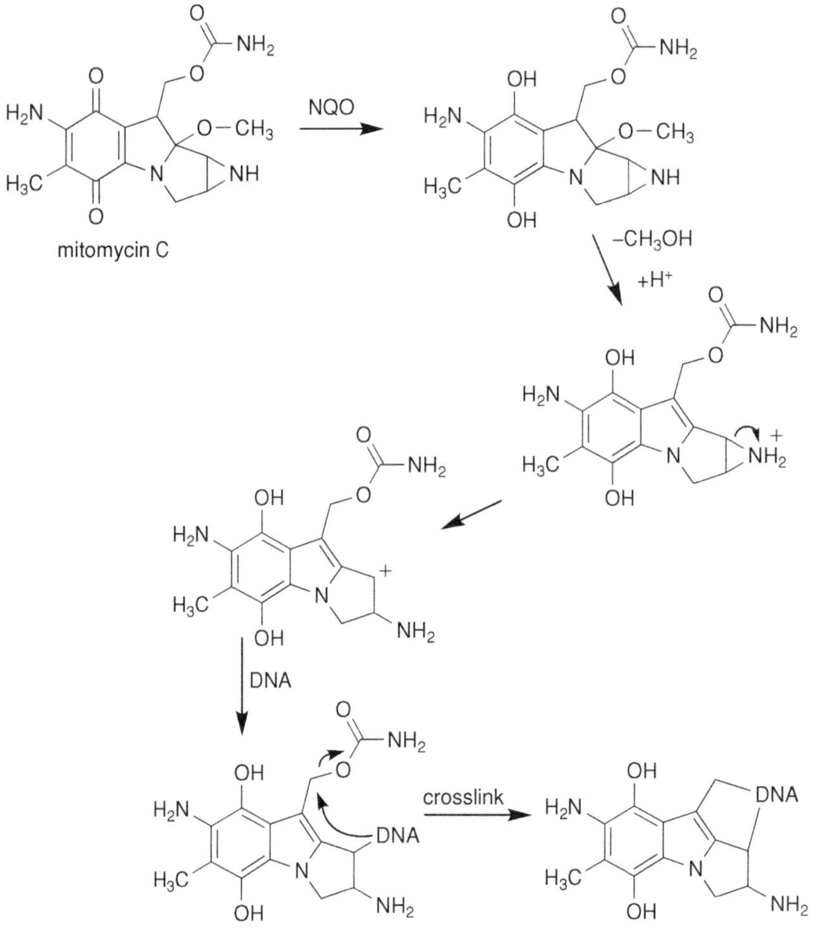

FIGURE 5.12 Bioactivation of mitomycin C, the first step of which involves reduction of the quinone.

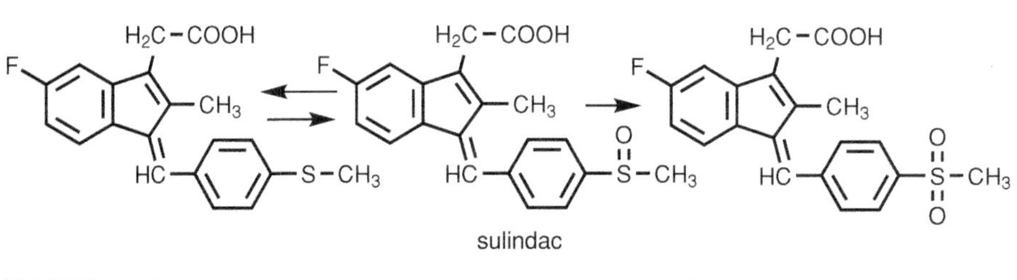

FIGURE 5.13 Reduction of the sulfoxide sulindac to the sulfide is reversible, but the sulfone metabolite is not reduced back to sulindac.

Reduction of Disulfides and Other Oxidation States of Sulfur

Disulfides can be reduced to two thiols (Fig. 5.14). The best example is the reduction of oxidized glutathione (GSSG) back to the reduced form (GSH) (Fig. 5.14), which is mediated by glutathione reductase. In addition, exchange can occur with other thiols mediated by protein disulfide isomerase. In principle, sulfenic acids can probably also be reduced back to thiols, but because of the reactivity of the sulfenic acid, this is not generally observed.

$$G-S-S-G \longrightarrow 2 \ G-SH$$

FIGURE 5.14 Disulfides such as oxidized glutathione (GSSG) are reduced to thiols (GSH).

Some other oxidation states of sulfur can be reduced but not sulfonic acids or, as mentioned above, sulfones.

Reduction of Peroxides

Analogous to the reduction of disulfides, peroxides and hydroperoxides, compounds that are generally toxic are readily reduced back to alcohols by peroxidases; however, in the process, other compounds including drugs can be oxidized.

Reductive Dehalogenation

A one-electron reduction of the bond between an aliphatic carbon and a halogen leads to a halogen anion and a carbon-free radical. A good example is the reduction of carbon tetrachloride as discussed earlier in this chapter. The first product in the reduction is the trichloromethyl-free radical. Carbon-centered radicals are not very reactive with biological molecules, but they react very rapidly with molecular oxygen (a diradical) to form a peroxy-free radical (Fig. 5.15), which is quite toxic (10).

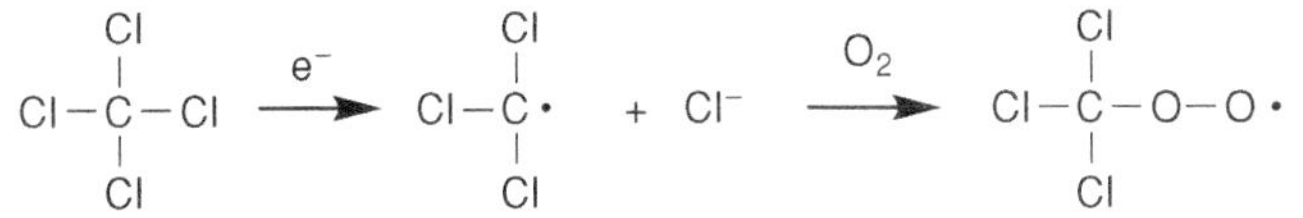

FIGURE 5.15 Reductive dehalogenation of carbon tetrachloride results in a carbon-centered free radical that reacts rapidly with oxygen to form the toxic peroxy radical.

The ease of reductive dehalogenation depends on the degree of electron deficiency of the molecule; halothane is somewhat more difficult to reduce than carbon tetrachloride and, in general, reductive dehalogenation does not occur with a monochloro compound. Most reductive dehalogenation reactions are likely mediated by cytochromes P450 in which the halogenated compound competes with molecular oxygen for reduction by the P450/P450 reductase; therefore, such reduction occurs more readily under hypoxic conditions (21).

REFERENCES

1. Matsunaga T, Shintani S, Hara A. Multiplicity of mammalian reductases for xenobiotic carbonyl compounds. Drug Metab Pharmacokinet 2006;21(1):1–18.
2. Martin HJ, Breyer-Pfaff U, Wsol V, et al. Purification and characterization of akr1b10 from human liver: role in carbonyl reduction of xenobiotics. Drug Metab Dispos 2006;34(3):464–470.
3. Breyer-Pfaff U, Nill K. Carbonyl reduction of naltrexone and dolasetron by oxidoreductases isolated from human liver cytosol. J Pharm Pharmacol 2004;56(12):1601–1606.
4. Schaller M, Schaffhauser M, Sans N, et al. Cloning and expression of succinic semialdehyde reductase from human brain. Identity with aflatoxin B1 aldehyde reductase. Eur J Biochem 1999;265(3):1056–1060.

5. Forrest GL, Gonzalez B. Carbonyl reductase. Chem Biol Interact 2000;129(1–2):21–40.

6. Ohara H, Miyabe Y, Deyashiki Y, et al. Reduction of drug ketones by dihydrodiol dehydrogenases, carbonyl reductase and aldehyde reductase of human liver. Biochem Pharmacol 1995;50(2):221–227.

7. Kaminsky LS, Zhang ZY. Human P450 metabolism of warfarin. Pharmacol Ther 1997;73(1):67–74.

8. Ahr HJ, King LJ, Nastainczyk W, et al. The mechanism of reductive dehalogenation of halothane by liver cytochrome P450. Biochem Pharmacol 1982;31(3):383–390.

9. Nastainczyk W, Ahr HJ, Ullrich V. The reductive metabolism of halogenated alkanes by liver microsomal cytochrome P450. Biochem Pharmacol 1982;31(3):391–396.

10. Mico BA, Pohl LR. Reductive oxygenation of carbon tetrachloride: trichloromethylperoxyl radical as a possible intermediate in the conversion of carbon tetrachloride to electrophilic chlorine. Arch Biochem Biophys 1983;225(2):596–609.

11. Masana M, de Toranzo EG, Castro JA. Reductive metabolism and activation of benznidazole. Biochem Pharmacol 1984;33(7):1041–1045.

12. Tatsumi K, Kitamura S, Narai N. Reductive metabolism of aromatic nitro compounds including carcinogens by rabbit liver preparations. Cancer Res 1986;46(3):1089–1093.

13. Kitamura S, Tatsumi K. Reduction of tertiary amine N-oxides by liver preparations: function of aldehyde oxidase as a major N-oxide reductase. Biochem Biophys Res Commun 1984;121(3):749–754.

14. Ross D. Quinone reductases multitasking in the metabolic world. Drug Metab Rev 2004;36(3–4):639–654.

15. Vella F, Ferry G, Delagrange P, et al. NRH:quinone reductase 2: an enzyme of surprises and mysteries. Biochem Pharmacol 2005;71(1–2):1–12.

16. Lee JM, Johnson JA. An important role of Nrf2-ARE pathway in the cellular defense mechanism. J Biochem Mol Biol 2004;37(2):139–143.

17. Smith KS, Smith PL, Heady TN, et al. In vitro metabolism of tolcapone to reactive intermediates: relevance to tolcapone liver toxicity. Chem Res Toxicol 2003;16(2):123–128.

18. Klotz U. Clinical pharmacokinetics of sulphasalazine, its metabolites and other prodrugs of 5-aminosalicylic acid. Clin Pharmacokinet 1985;10(4):285–302.

19. Hermans JJ, Thijssen HH. Properties and stereoselectivity of carbonyl reductases involved in the ketone reduction of warfarin and analogues. Adv Exp Med Biol 1993;328:351–360.

20. Suresh Kumar G, Lipman R, Cummings J, et al. Mitomycin C-DNA adducts generated by DT-diaphorase. Revised mechanism of the enzymatic reductive activation of mitomycin C. Biochemistry 1997;36(46):14128–14136.

21. Knights KM, Gourlay GK, Cousins MJ. Changes in rat hepatic microsomal mixed function oxidase activity following exposure to halothane under various oxygen concentrations. Biochem Pharmacol 1987;36(6):897–906.

6

Hydrolytic Pathways

Many drugs have functional groups that can be metabolized by the addition of water. The major functional groups involved are esters, amides, and epoxides. Several phase II metabolites such as sulfates and glucuronides, which will be discussed in Chapter 7, can also be hydrolyzed back to the parent drug.

HYDROLYSIS OF ESTERS, AMIDES, AND THIOESTERS

The relative ease of the hydrolysis is thioesters > esters > amides, and the products are a carboxylic acid and a thiol, alcohol, or amine, respectively, as shown in Figure 6.1. Amides are more difficult to hydrolyze than esters because the lone pair of electrons on the nitrogen is delocalized into the bond between the nitrogen and the carbonyl carbon leading to a bond order greater than one, i.e., it is more than a single bond (this is the same reason that amides are not basic as discussed in Chapter 2). This is more important in amides than esters because oxygen is more electronegative and therefore the nonbonded electrons of oxygen are less prone to delocalize than the lone pair of electrons of nitrogen. Although sulfur is even less electronegative, the lone pair of electrons is in an outer orbital that does not significantly overlap with the p orbitals of the carbonyl carbon.

A good example to illustrate the difference in the rates of hydrolysis of esters and amides is to compare the metabolism of procaine and procainamide because the only difference between the two drugs is that one is an ester and the other is an amide (Fig. 6.2). Procaine has a half-life of about 1 minute due to the rapid hydrolysis of the ester, whereas

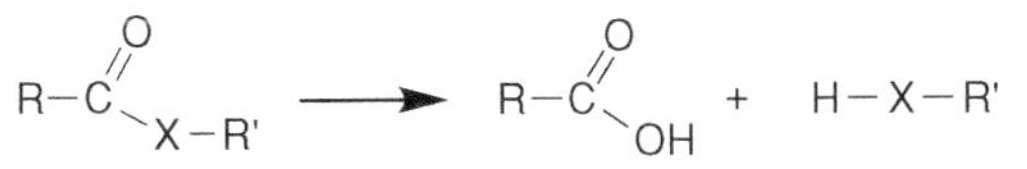

X = O (ester), N (amide), or S (thioester)   S > O > N

FIGURE 6.1 Hydrolysis of esters, thioesters, and amides.

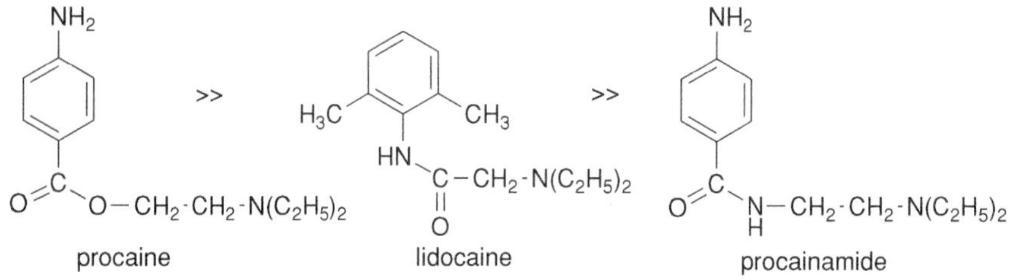

FIGURE 6.2 Examples of the relative rates of hydrolysis of an ester, the amide of an aromatic amine, and the amide of an aliphatic amine.

procainamide has a half-life of 4 hours with its major mode of clearance being renal rather than metabolic and very little p-aminobenzoic acid is observed as a metabolite. However, if the amide involves an aromatic amine, the effect of nitrogen lone pair of electrons that increase the bond order of the N-carbonyl bond and make it more difficult to hydrolyze an amide, as described above, is lessened because these electrons are also delocalized into the aromatic ring.

Another way of rationalizing the difference between amides involving aliphatic and aromatic amines is that in aromatic amines the nitrogen lone pair of electrons is delocalized into the aromatic ring and less available to satisfy the partial positive charge on the carbonyl carbon thus making the carbonyl group more electrophilic and susceptible to nucleophilic attack by the enzyme. In the case of aliphatic amides, the nitrogen lone pair is free to delocalize to the carbonyl carbon, satisfy the partial positive charge, and lessen susceptibility to nucleophilic attack. Ultimately, an aromatic system attached to the nitrogen shifts electron density away from the amide and the result is that amides involving an aromatic amine are, in general, hydrolyzed more rapidly than amides involving an aliphatic amine. It is also the reason why aromatic amines are weaker bases than aliphatic amines, as explained in Chapter 2. In contrast, an aromatic system attached to the carbonyl group does not have a significant effect on the rate of hydrolysis just as an aromatic ring does not significantly increase the acidity of a carboxylic acid as mentioned in Chapter 2. For example, lidocaine (Fig. 6.2) is more readily hydrolyzed than procainamide and dimethylaniline is a significant metabolite of lidocaine.

There are few drugs that are thioesters, but you may recall that one of the intermediates in the oxidation of aldehydes by aldehyde dehydrogenase is a thioester involving the thiol of the enzyme (Fig. 30 in Chapter 4), which is readily hydrolyzed back to the native form of the enzyme, a thiol, and the carboxylic acid product. Some drugs that are carboxylic acids, such as enaloprilate, are administered as ester prodrugs (enalopril), which are more readily absorbed from the intestine than the carboxylic acid and are then readily hydrolyzed to the active drug by esterases as mentioned in Chapter 1 (Fig. 1 in Chapter 1).

Although hydrolytic enzymes, esterases and amidases, are named after their major substrates, the same enzyme can often hydrolyze esters, thioesters, and amides; therefore, the differentiation between esterases and amidases is sometimes artificial. The highest hydrolytic activity is in the liver, but the enzyme pseudocholinesterase is found in the serum. Gut bacteria also contain hydrolytic enzymes.

Esterases

Esterases that contribute to human drug metabolism fall into three major classes: the cholinesterases (acetylcholinesterase, pseudocholinesterase, butyrylcholinesterase, etc.),

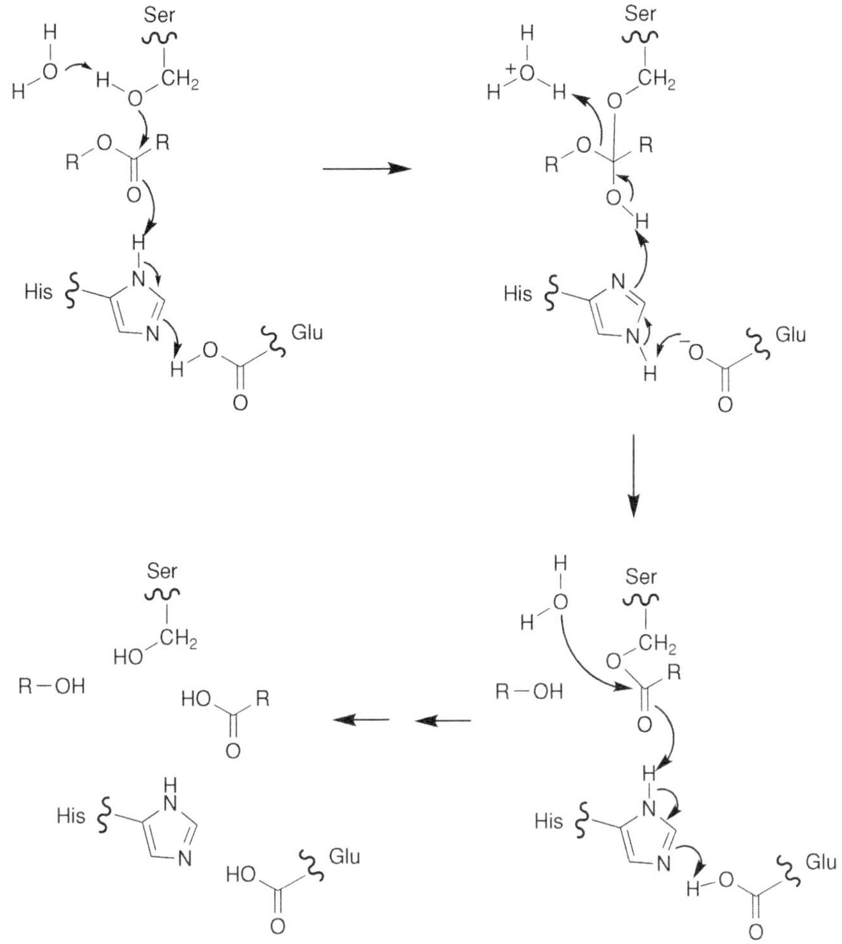

FIGURE 6.3 Mechanism for esterase-catalyzed hydrolysis of esters and amides.

carboxylesterase (CES), and paraoxonase. All the esterases operate using the same catalytic triad of Ser-His-Glu in which the hydroxyl group of Ser attacks the carbonyl group of the ester substrate (Fig. 6.3). His and Glu act as a charge relay system in which, in a synchronous step, a proton is transferred from Glu to N3 of His as the proton on NI of His protonates the developing negative charge on the ester carbonyl of the substrate arising from the attack of the Ser hydroxyl. The process lowers the activation energy for formation of the tetrahedral intermediate. Formation of the tetrahedral intermediate is followed by a reversal of electron flow from the hydroxyl group of the tetrahedral intermediate to N1 of His and transfer of the His N3 proton to Glu. This catalyzes bond cleavage, release of choline, and generation of the acetylated enzyme. Subsequent hydrolysis of the acetylated species regenerates active enzyme and acetic acid.

Acetylcholinesterase

The primary function of acetylcholinesterase is to terminate the activity of the neurotransmitter, acetylcholine (Fig. 6.4), through hydrolysis at the various cholinergic nerve endings. In this regard, it is probably the most highly efficient enzyme that operates in the human. It is capable of hydrolyzing 300,000 molecules of acetylcholine per molecule of enzyme

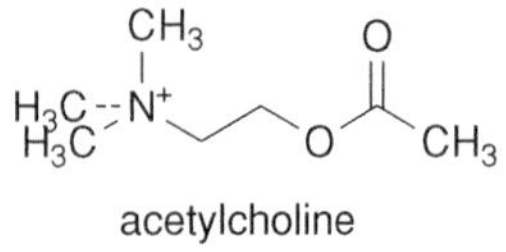

acetylcholine

FIGURE 6.4 Structure of acetylcholine.

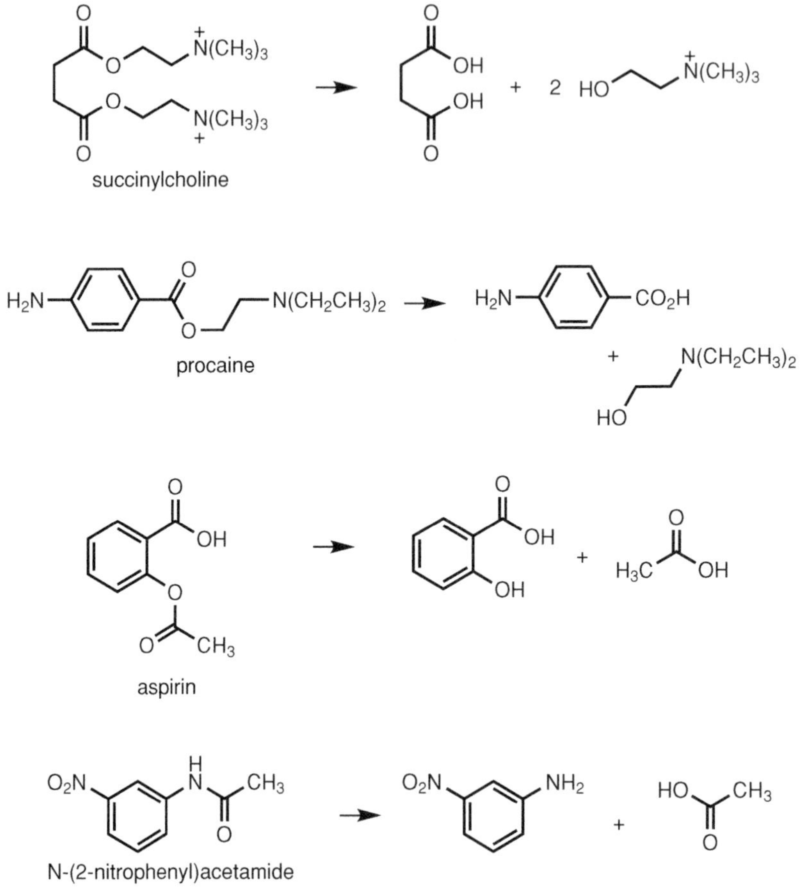

succinylcholine

procaine

aspirin

N-(2-nitrophenyl)acetamide

FIGURE 6.5 Substrates for pseudocholinesterase.

per minute. The basic unit is a homotetramer of 80-kDa subunits that is tethered to membrane (nerve endings) or cell surface (red blood cells) by a glycolipid anchor (1). While acetylcholinesterase is very efficient at hydrolyzing acetylcholine, it is not one of the major hydrolytic enzymes involved in drug metabolism.

Pseudocholinesterase

The contribution of pseudocholinesterase, also known simply as cholinesterase, to drug metabolism is much greater as it possesses considerably broader substrate selectivity. In addition to acetylcholine, it will hydrolyze other choline esters like the muscle relaxant succinylcholine. It will also hydrolyze non-choline-containing drugs like the local anesthetic procaine and the anti-inflammatory agent aspirin (Fig. 6.5). Cholinesterases, particularly butyrylcholinesterase, are also known to be able to catalyze the hydrolysis of arylamides, e.g., *N*-(2-nitrophenyl)acetamide, and alkyl (acid portion of the amide) analogs (2).

Pseudocholinesterase is a polymorphic enzyme. Succinylcholine is a paralyzing agent used during surgery to prevent muscle twitching. When succinylcholine is used in patients who are deficient in pseudocholinesterase, they wake up from the anesthetic but remain paralyzed for a prolonged period of time.

Carboxylesterases

In humans, two major forms of carboxylesterase (CES) have been identified, isozymes hCE-1 and hCE-2, that are members of the CES1 and CES2 families, respectively. They are smaller proteins than acetylcholinesterase as they have 60-kDa subunits rather than 80 kDa and their distribution is different. Both isozymes are found in multiple tissues (liver, heart, brain, testes, etc.), but hCE-1 is located primarily in the liver while hCE-2 is found primarily in the small intestine. Both are major contributors to drug metabolism. However, the two enzymes differ in substrate selectivity: hCE-1 tends to hydrolyze substrates with a large acyl group but a small alcohol group, whereas the exact opposite is true of hCE-2—it tends to hydrolyze substrates that have a large alcohol group but a smaller acyl group. For example, both enzymes hydrolyze cocaine, but hCE-1 hydrolyzes the methyl ester structural unit while hCE-2 hydrolyzes the benzoyl ester structural unit (Fig. 6.6). Other hCE-1 substrates include the analeptic and anorexic agent methylphenidate and the nonsteroidal anti-inflammatory agent flurbiprofen ethyleneglycol, while other hCE-2 substrates include heroin and *p*-nitrophenylacetate (3).

While there are clear differences in substrate selectivity between the drug metabolizing hydrolytic enzymes, there is also significant overlap, i.e., they will often tend to metabolize the same substrates but at different rates. For example, pseudocholinesterase, hCE-1, and hCE-2 catalyze the hydrolysis of heroin and cocaine.

The introduction of an ester function into a hydroxyl group–containing therapeutic agent that is orally administered generally increases the drug's bioavailability through an increase in absorption. This knowledge coupled to the knowledge of the catalytic and distributional properties of the various hydrolases has been very useful in developing prodrugs of poorly absorbed drugs.

Paraoxonase

Paraoxonase (PON1) is a 43-kDa serum protein almost exclusively associated with high-density lipoprotein. Initially, PON1 was identified as an enzyme that would hydrolyze and deactivate paraoxon (Fig. 6.7), the active metabolite of parathion), hence the name. It also hydrolyzes organophosphates in general. This was a particularly significant discovery because paraoxon is a suicide substrate inhibitor of both pseudocholinesterase in serum and acetylcholinesterase in serum, at synapses, and the neuromuscular junction. Thus, PON1 appears to be the body's main defense mechanism against the potentially lethal neurotoxicity of organophosphates that might result upon exposure to the toxin. In addition to its ability to hydrolyze organophosphates, PON1 appears to have a major protective role in suppressing the development of atherosclerosis promoted by the oxidation of low-density lipoprotein (4).

Amidases

Given the overwhelming occurrence of the amide bond in the food we eat and the proteins, peptides, and enzymes that are large components of the structural and catalytic elements of

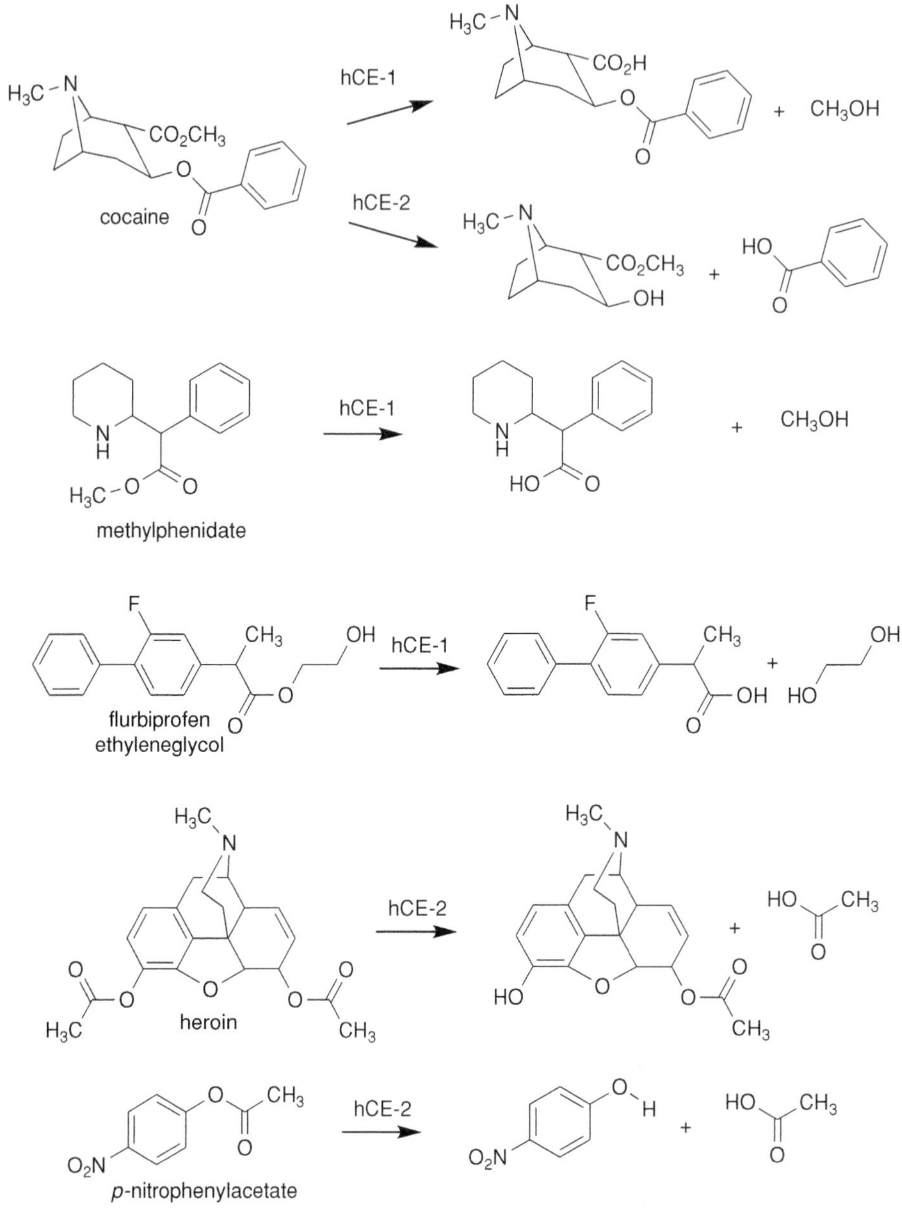

FIGURE 6.6 The selectivity of hCE1 and hCE2 for catalyzing the hydrolysis of various ester substrates.

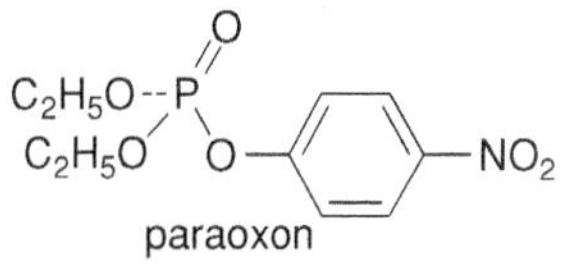

FIGURE 6.7 Structure of paraoxon.

our bodies, it is not surprising that the body contains numerous enzymes that are capable of breaking that bond; a process that occurs through hydrolysis. However, in an in vivo environment, xenobiotic amides tend to be more stable than esters to hydrolysis just as they are more stable in a strictly chemical acid or base environment as discussed previously.

A number of proteolytic enzymes are excreted into the digestive tract (stomach and small intestine) to affect the hydrolysis of ingested proteins and break them down into their constitutive amino acids. These enzymes are categorized as either endopeptidases or exopeptidases; the former hydrolyze internal amide bonds while the later hydrolyze terminal amide bonds. Endopeptidases include trypsin, an enzyme that is particularly active toward amino acid residues in which the amino component is contributed by aromatic and acidic amino acid residues, chymotrypsin that hydrolyzes amino acids whose carbonyl component contains the aromatic amino acid residues and to a lesser extent Leu and Met, and elastase, an enzyme that activates amide bonds involving the neutral amino acids. The exopeptidases include carboxypeptidase A, a zinc-containing enzyme that hydolyzes all carboxyl terminal linkages except for Lys, Arg, or if the penultimate residue is Pro. Carboxypeptidase B is also a zinc-containing exopeptidase, which is complimentary to carboxypeptidase A in that it will only hydrolyze carboxyl terminal Lys or Arg residues. Another exopeptidase is leucine aminopeptidase, also a zinc-containing enzyme, which despite its name is nonspecific and will hydrolyze most amino terminal peptide bonds. Thus, the appropriate machinery is released into the gut to efficiently degrade a protein to its constituent amino acids that are then actively absorbed.

In terms of drug metabolism, it appears that the aminopeptidases, along with the cholinesterases as discussed above, are probably the most efficient in hydrolyzing the amide bonds of drugs.

HYDROLYSIS OF EPOXIDES

Epoxides are compounds that contain the chemically reactive structural element of a highly strained three-membered oxygen-containing ring. They can be formed in vivo by the cytochrome P450–catalyzed oxidation of a carbon–carbon double bond (see discussion in "Alkenes" and "Aromatic Rings" sections in Chapter 4). Because they are susceptible to attack by endogenous nucleophiles, e.g., sulfhydryl, amino, and hydroxyl groups, which can lead to covalent bond formation resulting from the inactivation and modification of critical biomacromolecules, they tend to be toxic. In fact, the *trans*-7,8-dihydrodiol-9,10-epoxy metabolic product of the environmental contaminant, benzo[*a*]pyrene, is highly carcinogenic (see Chapter 8).

Epoxides/arene oxides have varying degrees of chemical reactivity and can be detoxified by hydrolysis to dihydrodiols as shown in Figure 6.8. This can occur either nonenzymatically, if the epoxide is very reactive, or it can be catalyzed enzymatically by epoxide hydrolase (EH).

The EHs are a family of enzymes that deactivate epoxides by catalyzing their hydration to form diols (5) (Fig. 6.9). Two members of this family are associated with drug

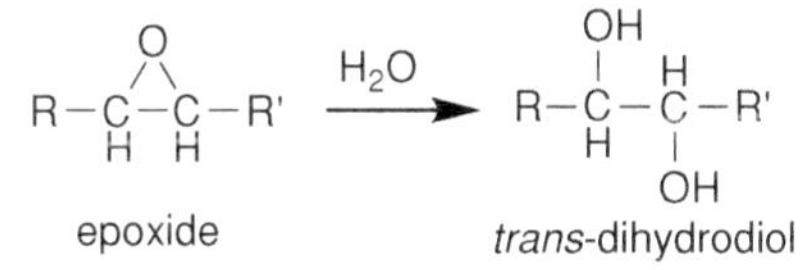

FIGURE 6.8 Hydrolysis of epoxides to *trans*-dihydrodiol.

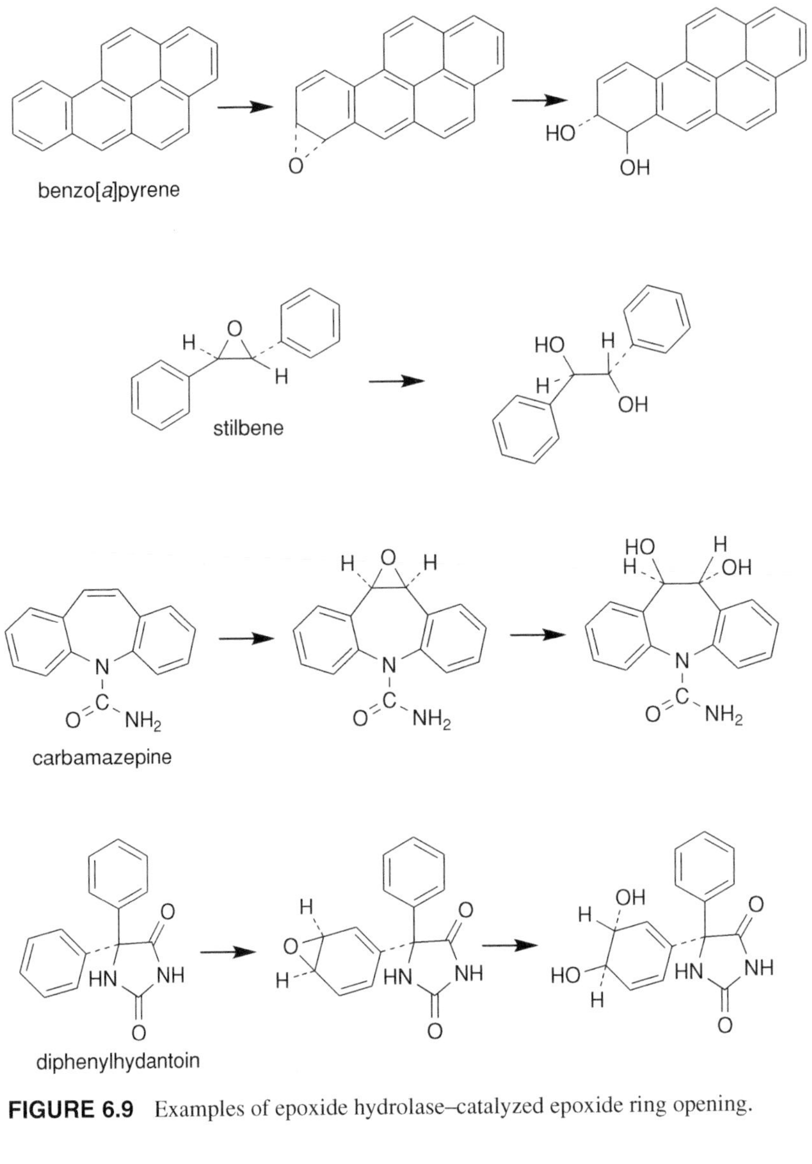

FIGURE 6.9 Examples of epoxide hydrolase–catalyzed epoxide ring opening.

metabolism, a soluble form and a membrane-bound microsomal form. The soluble form of EH is a 62-kDa protein that is expressed in virtually all tissues and is found in cytosol. The microsomal form of EH is a 49-kDa protein which is also expressed in virtually all tissues localized in the endoplasmic reticulum. Soluble EH is characterized by the hydration of *trans* stilbene oxide, although its primary role appears to be more physiological and is related to the hydration of endogenous epoxides such as the epoxides of arachidonic acid.

In contrast, the primary role of microsomal EH appears to be in detoxifying the metabolically produced epoxides of drugs, e.g., carbamazepine epoxide, the arene oxide of diphenylhydantoin, and the epoxides of environmental contaminants like the polycyclic aromatic hydrocarbons, e.g., benzo[a]pyrene.

Ring opening of the epoxide and generation of the diol product proceeds by a two-step process. In the first step the epoxide is opened forming an alkylated enzyme, and in the second step the diol product is released and enzyme is regenerated. Mechanistically, the two

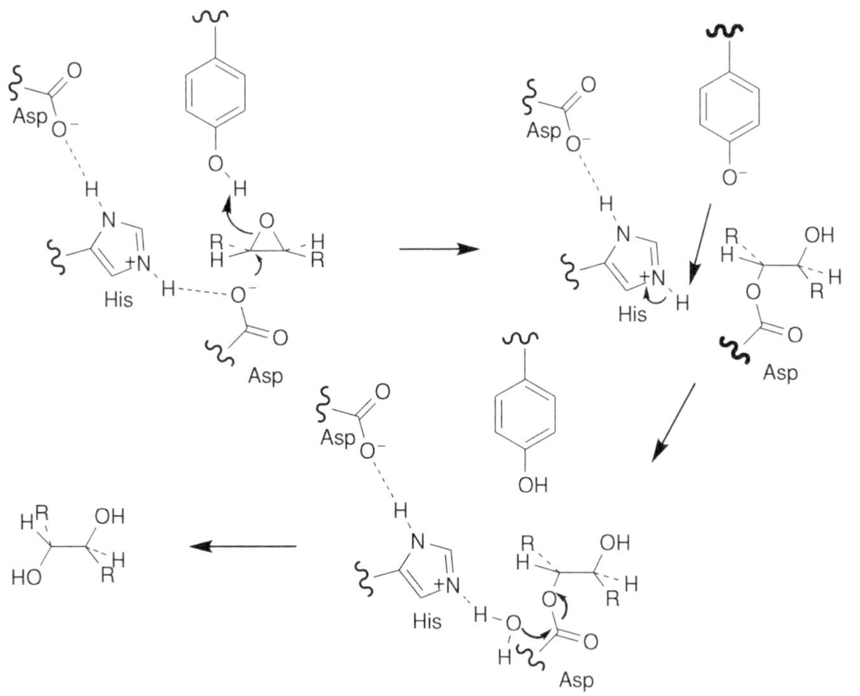

FIGURE 6.10 Mechanism of epoxide hydrolase–catalyzed hydrolytic opening of an epoxide.

steps involve the action of a catalytic triad composed of Asp226, His431, and Glu404, in the case of human microsomal EH, and Asp334, His523, and Asp495, for human soluble EH (6) (Fig. 6.10). Reaction is initiated by nucleophilic attack of an Asp carboxyl (Asp226 or Asp334) on a carbon atom of the epoxide ring. The nucleophilicity of the attacking Asp is increased by His removing the Asp proton. Breaking of the carbon–oxygen bond to open the ring is further assisted by Tyr residues (Fig. 6.10) that hydrogen bond to the epoxide oxygen to stabilize and neutralize the developing negative charge. Completion of ring opening and proton transfer leads to formation of the enzyme alkyl intermediate. His transfers a proton to the Tyr anion, and then activates a water molecule to hydrolyze the intermediate and release diol. The Glu404 and Asp495 residues serve as orienting species.

The observation of a dihydrodiol has been taken as proof that an epoxide (arene oxide) is the precursor metabolite. Many epoxides, such as the 10,11-epoxide of carbamazepine shown above and even the arene oxide of benzene, which is quite reactive, have been directly observed. Others such as the epoxide of phenytoin are only inferred. It is conceivable that some dihydrodiols are formed by reaction of an intermediate with water in the active site of P450 without the formation of an epoxide. One clue to the origin of the dihydrodiol is the stereochemistry; an exclusively *trans*-dihydrodiol suggests that it was formed via the EH-mediated hydrolysis of an epoxide or arene oxide.

HYDROLYSIS OF SULFATES AND GLUCURONIDES

Many drugs and metabolites are metabolized by conjugation with sulfate or glucuronic acid as described in Chapter 7. Sulfate conjugates can be hydrolyzed back to the alcohol or phenol. Glucuronide conjugates can involve a wider variety of functional groups and many of these conjugates can also be hydrolyzed back to the parent drugs. These hydrolytic pathways generally occur in the intestine and are mediated either by bacterial enzymes or intestinal enzymes. Such hydrolysis can lead to enterohepatic cycling where a drug is conjugated with either sulfate or glucuronic acid in the liver, excreted into bile, and then hydrolyzed back to the parent drug in the intestine thus permitting reabsorption through the portal system into the liver where the drug can undergo conjugation again to complete the cycle. Treatment with antibiotics can decrease enterohepatic cycling by decreasing the bacterial contribution to hydrolysis and this may decrease levels of drugs such as contraceptive steroids.

REFERENCES

1. Taylor P. Anticholinesterase agents. In: Gilman AG, Goodman LS, Gilman A, eds. The Pharmacologic Basis of Therapeutics (6th ed). New York, NY: Macmillan Publishing Co.; 1980.
2. Darvesh S, McDonald RS, Darvesh KV, et al. On the active site for hydrolysis of aryl amides and choline esters by human cholinesterases. Bioorg Med Chem 2006;14(13):4586–4599.
3. Imai T. Human carboxylesterase isozymes: catalytic properties and rational drug design. Drug Metab Pharmacokinet 2006;21(3):173–185.
4. Durrington PN, Mackness B, Mackness MI. Paraoxonase and atherosclerosis. Arterioscler Thromb Vasc Biol 2001;21(4):473–480.
5. Omiecinski CJ. Epoxide hydroxylases. In: Levy RH, Thummel KE, Trager WF, et al., eds. Metabolic Drug Interactions. Philadelphia, PA: Lippincott, Williams & Wilkins; 2000.
6. Morisseau C, Hammock BD. Epoxide hydrolases: mechanisms, inhibitor designs, and biological roles. Annu Rev Pharmacol Toxicol 2005;45:311–333.

7
Conjugation Pathways

The second major class of enzymes that serve to protect the organism from the potential toxicity of foreign compounds are the conjugating enzymes. In general, conjugation pathways involve the addition of a hydrophilic group such as glucuronic acid (see below) to a drug, and the mechanism involves an enzyme and a cofactor that is the source of this hydrophilic group. The cofactor usually contains a high-energy bond, such as a diphosphate, that facilitates the reaction. Conjugation usually adds a charge to the drug thus making the drug more polar and facilitating renal excretion; however, some conjugation pathways, in particular methylation and acetylation, do not increase the polarity of the substrate, but these two pathways do usually decrease pharmacological activity.

GLUCURONIDATION

Glucuronidation is the most common conjugation pathway, both because of the range of substrates that can undergo glucuronidation and because it is often a quantitatively important pathway. The enzymes involved are called glucuronosyl transferases and the cofactor is uridine-5'-diphospho-α-D-glucuronic acid (UDPGA). The mechanism involves nucleophilic attack of the substrate on the cofactor (S_N2 reaction) and leads to inversion of the configuration of the carbon involved thus converting the α-glucuronic acid to a β-glucuronide (Fig. 7.1). The reaction is shown with an oxygen nucleophile as the substrate (R–OH) but other nucleophilic atoms can make a drug a substrate for glucuronidation, i.e., nitrogen, sulfur, or even carbon (carbon anions) (1) leading to O-, N-, S-, or C-glucuronides, respectively.

The most common substrates for glucuronidation have an OH group, i.e., alcohols, phenols, and carboxylic acids that lead to the formation of O-glucuronides (2). The glucuronides of alcohols and phenols are called ether glucuronides because an ether linkage is formed between the drug and glucuronic acid, whereas the glucuronides of carboxylic acids are called ester glucuronides because the link between drug and glucuronic acid is an ester. N-glucuronides can be formed from primary aromatic amines, tertiary

uridine-5′-diphospho-a-D-glucuronic acid (UDPGA)   β-glucuronide

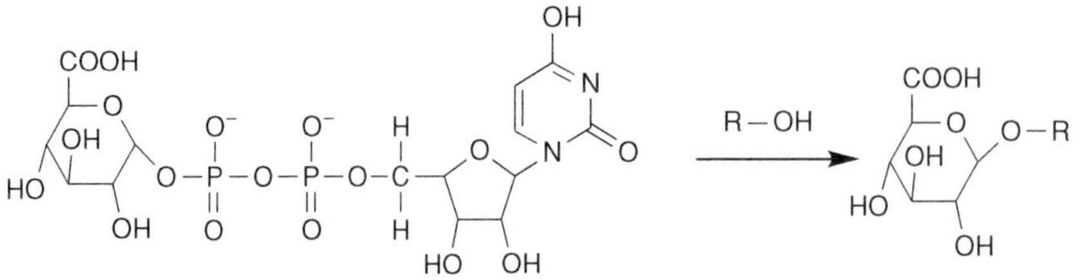

FIGURE 7.1 Glucuronidation of a substrate containing an OH group utilizing UDPGA as a cofactor.

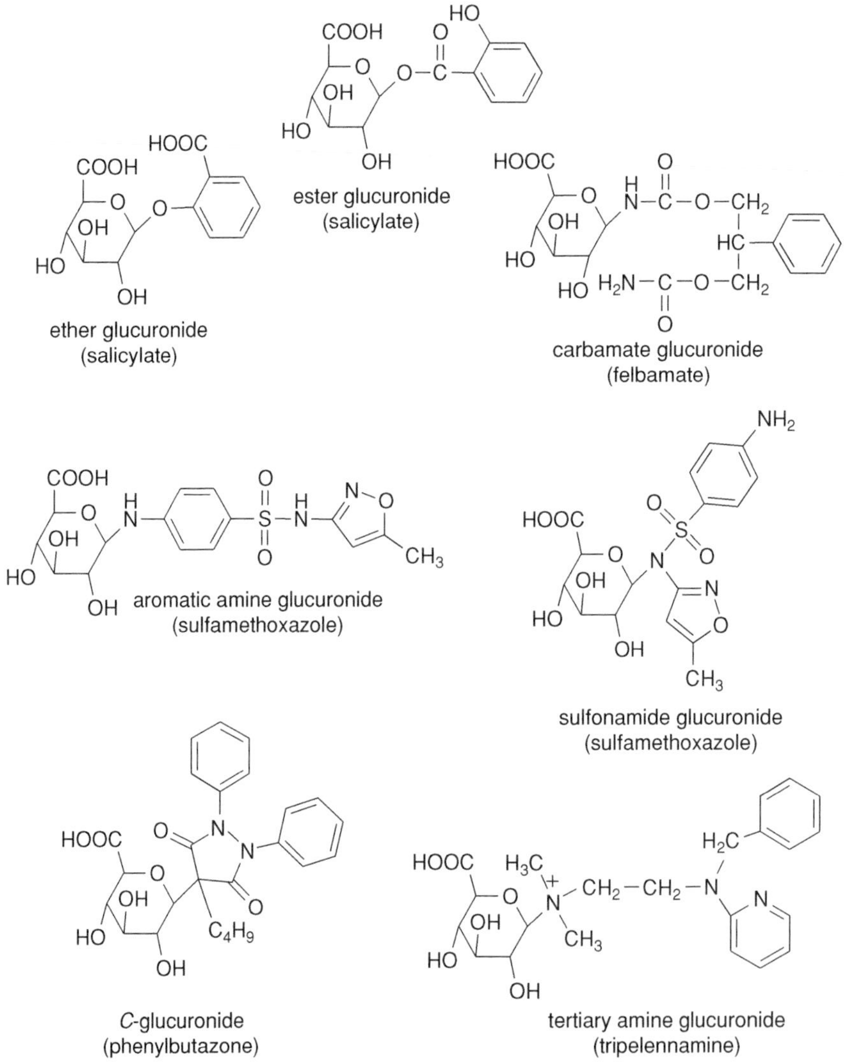

ester glucuronide
(salicylate)

ether glucuronide
(salicylate)

carbamate glucuronide
(felbamate)

aromatic amine glucuronide
(sulfamethoxazole)

sulfonamide glucuronide
(sulfamethoxazole)

C-glucuronide
(phenylbutazone)

tertiary amine glucuronide
(tripelennamine)

FIGURE 7.2 Examples of various types of glucuronide conjugates.

aliphatic amines, carbamates, and sulfonamides (3); *S*-glucuronides can be formed from thiophenols; and *C*-glucuronides can be formed from drugs that possess relatively acidic carbons such as phenylbutazone (4). Examples of various types of glucuronides are shown in Figure 7.2.

Glucuronosyl Transferases

The glucuronosyl transferases (UDPGTs) are a group of enzymes that belong to one of two gene families (UGT1 or UGT2) that catalyze the reaction described above, i.e., the S_N2 displacement of the uridine diphosphate group from UDPG by an attacking nucleophilic group from the compound that then becomes covalently bound to glucuronic acid forming the conjugated product (5). At least 16 different isozymes have been identified, cloned, and expressed: nine members of the UGT1 family (1A1, 1A3, 1A4, 1A5, 1A6, 1A7, 1A8, 1A9, and 1A10) and seven members of the UGT2 family (2B4, 2B7, 2B10, 2B11, 2B15, 2B17, and 2B28) (1). In general, the different isozymes can be viewed as having broad and overlapping substrate selectivity that vary in the efficiency, Vmax/Km,

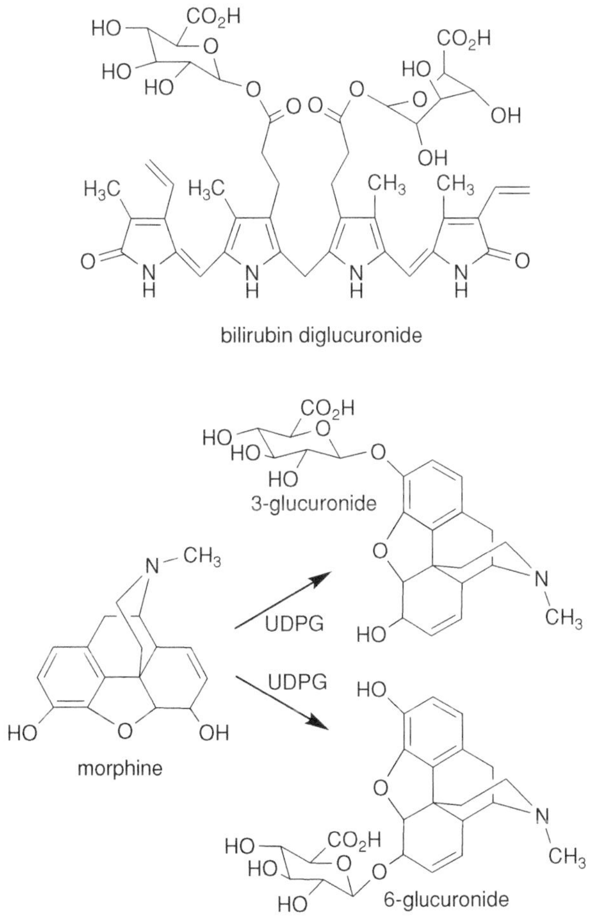

FIGURE 7.3 Structures of bilirubin diglucuronide and the 3- and 6-glucuronides of morphine.

with which they operate on various substrates. There are however several notable exceptions, e.g., UGT1A1—probably the most extensively studied human isozyme —is thought to be the only UGT that is physiologically relevant to the conjugation and elimination of bilirubin (Fig. 7.3), UGT1A9 has been found to be the only isozyme that is able to catalyze the C-glucuronidation of phenylbutazone (1), and UGT2B7 is particularly active in catalyzing the glucuronidation of opioids such as morphine. In the case of morphine, while the 3-*O*-glucuronide is the favored product, the 6-*O*-glucuronide is also formed.

Glucuronidation Characteristics

Glucuronidation is usually a low-affinity, high-capacity system. Therefore, if there are two competing metabolic pathways for a functional group, at low concentrations of the drug, the competing pathway will more likely be dominant whereas at high concentrations of the drug, glucuronidation will likely be dominant.

The activity of glucuronidation is low in the newborn, especially in premature babies (6). This is evident in the jaundice observed in many newborns because the major clearance pathway for bilirubin is glucuronidation. This can also lead to increased toxicity of some drugs in the newborn such as the "grey baby" syndrome seen in newborns treated with chloramphenicol.

As would be expected, glucuronidation, which adds a large negatively charged group to a drug, usually leads to a loss of the drug's pharmacological activity. However, there are exceptions; the affinity of the 6-glucuronide of morphine for the opiate receptor is approximately 100-fold greater than that of morphine itself, whereas the 3-glucuronide is an opiate antagonist (7). Glucuronides are also quite polar and do not usually penetrate the blood–brain barrier; however, the 6-glucuronide of morphine folds on to itself making it less polar and it readily penetrates the blood–brain barrier.

Ester glucuronides are somewhat chemically reactive and can covalently bind to protein. It has been proposed that this type of metabolite is responsible for the idiosyncratic reactions associated with several drugs that are carboxylic acids as discussed in Chapter 8.

SULFATION

The sulfotransferases are an emerging superfamily of enzymes that catalyze the transfer of SO_3^- to hydroxy or phenolic groups of susceptible substrates, the nitrogen of N-substituted aryl and alicyclic compounds, or pyridine N-oxides, through the action of the sulfating cofactor, 3'-phosphoadenosine 5'-phosphosulfate (Fig. 7.4) (8). Unlike glucuronidation, carboxylic acids are not substrates, and even if a conjugate were formed, it is unlikely that the product would be stable in an aqueous environment. Physiologically, they are involved in the biotransformation and elimination of steroid hormones and neurotransmitters (9). In contrast to glucuronidation, sulfation is usually low capacity, high affinity; therefore, with a substrate such as a phenol that can be either glucuronidated or sulfated, it is likely that sulfation will dominate at low concentrations whereas glucuronidation will dominate at high concentrations. While their substrate profile is limited relative to the UGTSs, the soluble sulfotransferases do have an important role. Sulfotransferases are widely distributed in the body and can lead to the formation of reactive metabolites in tissues such as skin that have low P450 activity.

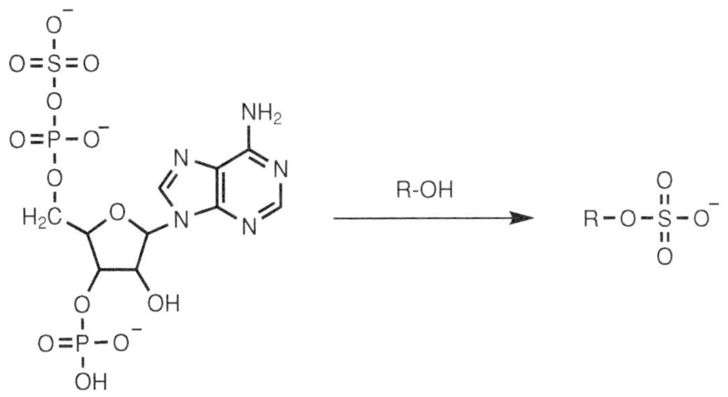

3'-phosphoadenosine-5'-phosphosulfate (PAPS)

FIGURE 7.4 Sulfation of a substrate containing an OH group utilizing PAPS as a cofactor.

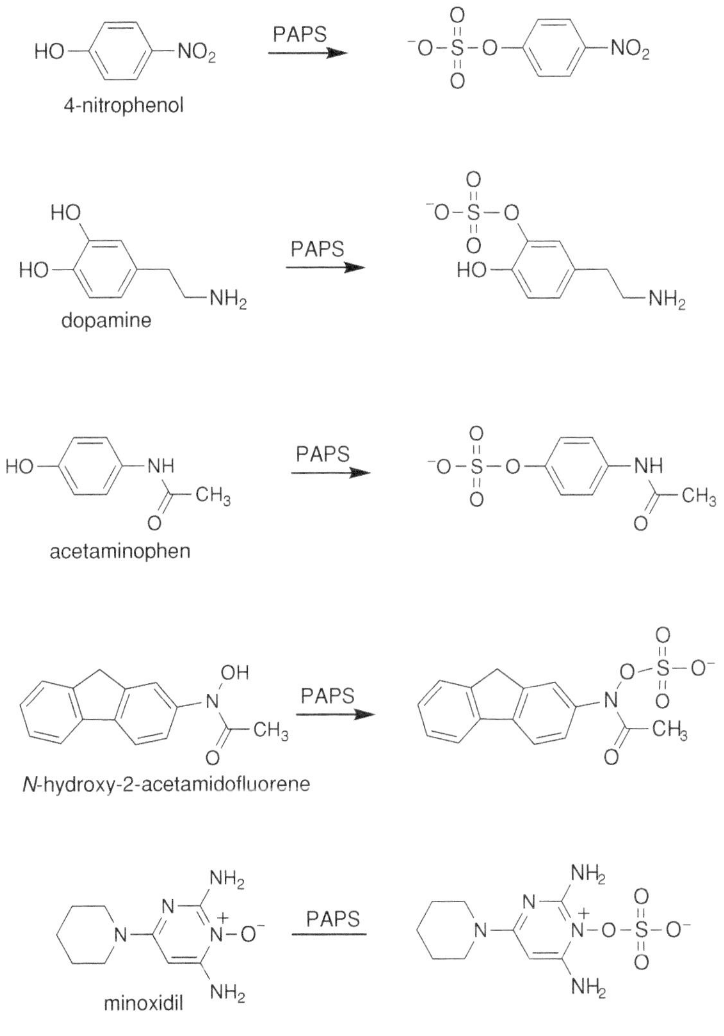

FIGURE 7.5 Examples of substrates for sulfotran sferases.

There are at least seven human isozymes known that are members of two different sulfotransferase families and five different subfamilies, SULT1A1, SULT1A21, SULT1A3, SULT1B1, SULT1C1, SULT1E1, and SULT2A1. Out of the seven known SULTs, the ones most likely to be involved in drug metabolism are the three SULT1A proteins. The prototypic substrates for these isozymes are 4-nitrophenol (1A1 and 1A2) or dopamine (1A3) (Fig. 7.5).

Two of the more well-known metabolic transformations catalyzed by the SULTs are the sulfation of the over-the-counter analgesic, acetaminophen, and the sulfation of the procarcinogen, N-hydroxy-2-acetamidofluorene (Fig. 7.5). Sulfate is a good leaving group, and therefore, the formation of a sulfate conjugate can lead to a reactive metabolite as discussed in Chapter 8. One interesting example of sulfation is the bioactivation of minoxidil, an agent that stimulates hair growth in the scalp, involving sulfation of the N-oxide in hair follicles (10). Thus, although sulfation usually leads to inactivation of the substrate, like glucuronidation, it can result in activation.

ACETYLATION

The major substrates for acetylation are primary aromatic amines, hydroxylamines (both the oxygen and the nitrogen can be acetylated), and hydrazines (11). The cofactor is acetyl Co-A, which is a thioester (Fig. 7.6).

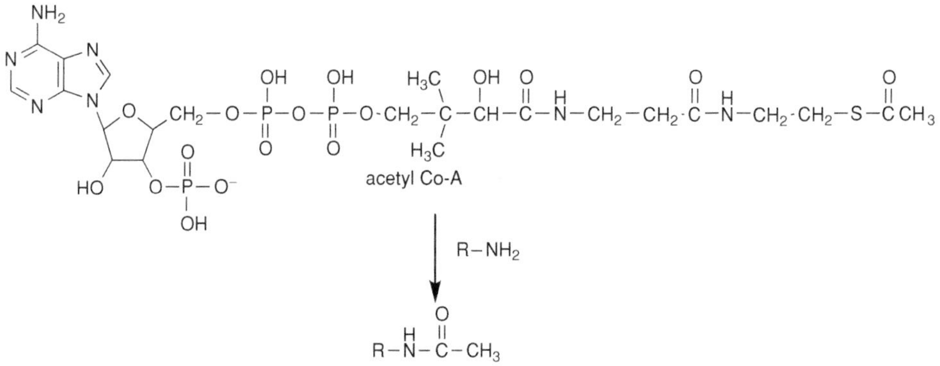

FIGURE 7.6 Acetylation of a substrate containing an NH_2 group utilizing acetyl Co-A as a cofactor.

The enzyme can also catalyze the transfer of an acetyl group from an N-acetylated hydroxylamine (hydroxamic acid) to form an acetoxy product, i.e., an N to O transacetylation and this pathway does not require acetyl Co-A (12). N-hydroxy-4-acetylaminobiphenyl provides an example of this conversion as shown in Figure 7.7. The significance of this pathway is that it leads to the activation of the hydroxamic acid because acetoxy derivatives of aromatic amines are chemically reactive and many are carcinogens such as the heterocyclic amines formed when meat is heated to a high temperature, e.g., 2-amino-1-methyl-6-phenylimidazo[4,5-b]pyridine.

There are two major N-acetyltransferase enzymes, NAT1 and NAT2 (13). Both enzymes have a cysteine at the active site, which is acetylated by acetyl Co-A, and NAT transfers the acetyl group to the substrate (ping-pong mechanism) (14). The two enzymes have overlapping specificity, but NAT1 prefers acidic substrates such as p-aminobenzoic acid and p-aminosalicylic acid whereas NAT2 is specific for sulfamethazine, hydralazine, and isoniazid (15) (Fig. 7.8). Steric hindrance caused by substituents ortho to the amino

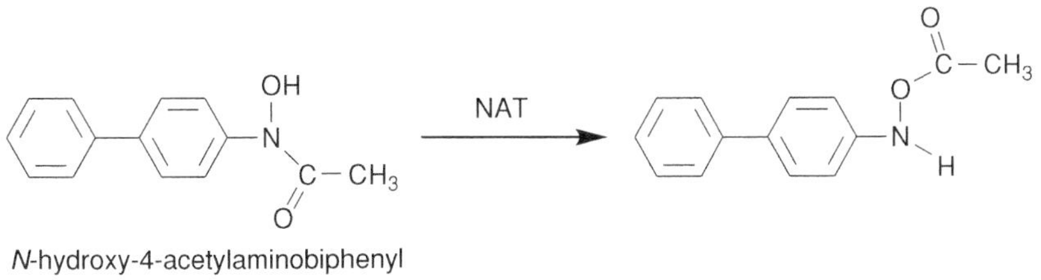

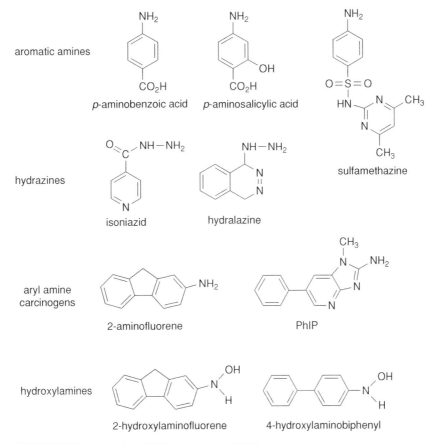

N-hydroxy-4-acetylaminobiphenyl

FIGURE 7.7 Transacetylation of *N*-hydroxy-4-acetylaminobiphenyl which converts it from a hydroxamic acid to the more reactive *N*-acetoxy-4-aminobiphenyl.

FIGURE 7.8 Examples of different types of NAT substrates.

group of aromatic amines usually prevents compounds from being good substrates for NAT1 (16). NAT1 activity is widely distributed, whereas NAT2 activity is quite high in the liver and there is also significant activity in the intestine but activity is lower outside of these two locations (17).

    Acetylation mediated by NAT2 was one of the first metabolic pathways found to be polymorphic. The rapid and slow phenotypes are approximately equal in prevalence in North America, whereas about 90% of Orientals exhibit the rapid acetylator phenotype while the ratio is reversed in people of Middle Eastern decent. Although the *NAT1* gene is also polymorphic, the corresponding phenotype is not as clear.

In general, acetylation decreases the polarity of a drug rather than increasing it, but the substrates, e.g., aromatic amines, hydroxylamines, and hydrazines, are often toxic, and acetylation usually decreases this toxicity. For example, the risk of neurotoxicity due to isoniazid and the risk of hydralazine-induced lupus are significantly higher in slow acetylators. The same is true for isoniazid-induced hepatotoxicity and procainamide-induced lupus, but the difference in risk between the two NAT2 phenotypes is smaller. The risk of certain cancers, especially bladder cancer, is also higher in patients of the slow NAT2 phenotype (18), but because NAT also catalyzes N to O transacetylation, acetylation can actually increase the carcinogenic potential of some compounds such as heterocyclic amines and benzidine.

METHYLATION

The methyltransferases represent a relatively large number of enzymes that utilize the cofactor, S-adenosyl-L-methionine, in which the methyl group is bound to a positively charged sulfur, to transfer a methyl group to an oxygen, sulfur, or nitrogen atom in an appropriate substrate as shown in Figure 7.9 (8).

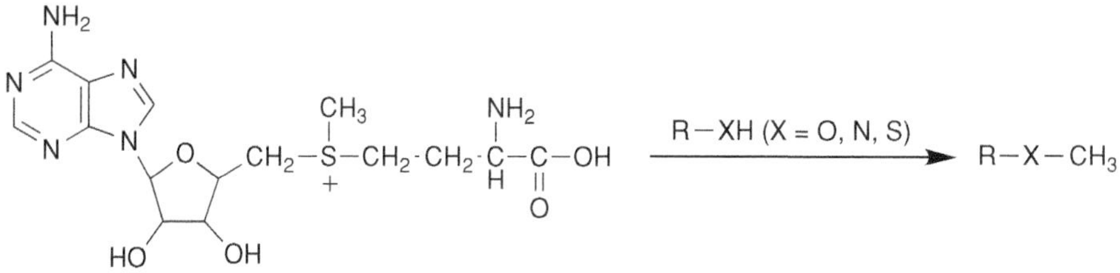

S-adenosylmethionine (SAM)

FIGURE 7.9 Reaction scheme for the methylation of substrates utilizing SAM as the cofactor.

Appropriate substrates include selected catechols, sulfhydryl, and nitrogen-containing compounds. Many of these compounds are endogenous such as histamine and the catechol neurotransmitters, e.g., norepinephrine (Figs. 7.10 and 7.11). Drugs, particularly those similar in structure to endogenous compounds that are substrates, are also prone to metabolic turnover by these enzymes. Catechol-O-methyltransferase (COMT), the phenolic methylating enzyme(s), have a molecular weight of approximately 25 kDa and come in both soluble and membrane-bound forms. The soluble form(s) is found in kidney and liver cytosol, while the membrane-bound form(s) has been reported to be localized in the brain. COMT has an absolute requirement for the catechol structure to be catalytically active. Thus, the scope of drugs it will metabolize is limited. Examples include norepinephrine, the anti-Parkinson's drug, L-dopa, and the β-agonist, isoproterenol (Fig. 7.10).

The number of drugs susceptible to S-methylation is still limited but greater than the number turned over by COMT. Thiopurine methyl transferase (TPMT) is an important enzyme responsible for detoxifying mercaptopurine—a drug used to treat leukemia—as well as azathioprine —a prodrug that is metabolized to mercaptopurine (Fig. 7.12). This enzyme is polymorphic and patients who are homozygous for the deficient enzyme experience severe toxicity when given usual doses of mercaptopurine (19). Similar aromatic and heterocyclic sulfhydryls can also be substrates for TPMT. The similar thiol

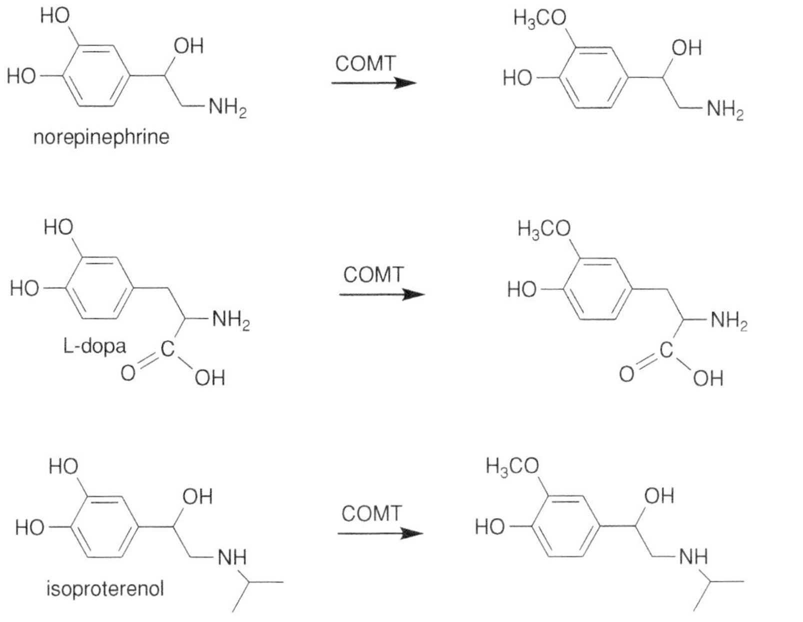

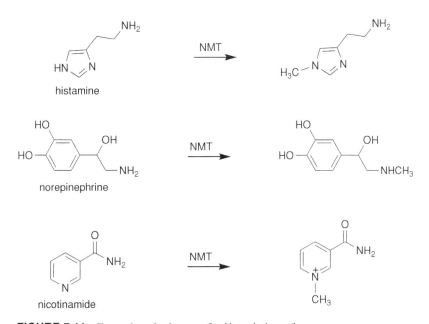

FIGURE 7.10 Examples of substrates for catechol-*O*-methyltransferase.

FIGURE 7.11 Examples of substrates for *N*-methyltransferases.

methyltransferase is responsible for the methylation of the aliphatic antihypertensive, captopril, and the anti-inflammatory agent, D-penicillamine (Fig. 7.12).

N-methyltransferase, active toward histamine and the catechol neurotransmitters, e.g., norepinephrine, is even more restrictive than COMT in terms of the metabolism of exogenous compounds. A class of compounds that does appear to be susceptible to N-methylation are azaheterocycles, particularly those that contain pyridine as part of the

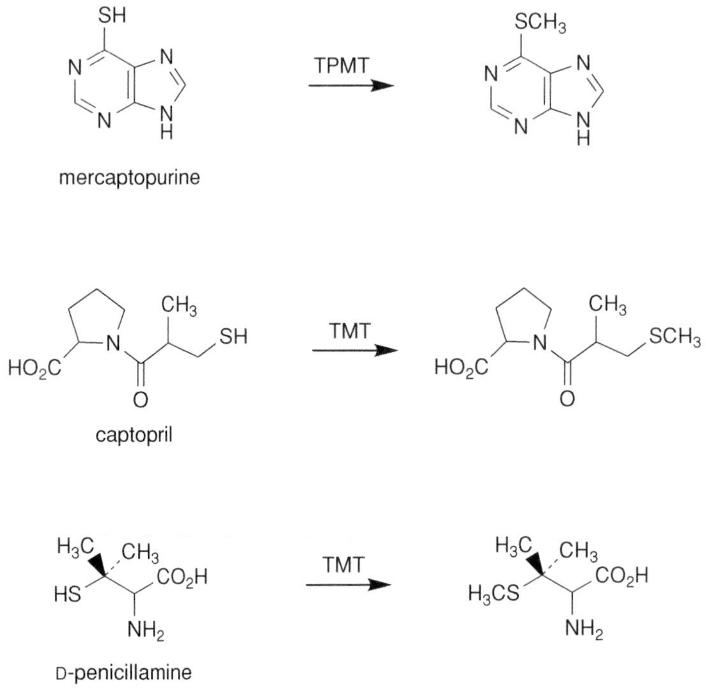

FIGURE 7.12 Examples of thiols that are substrates for methyltransferases.

structure (20). A well-known example is the N-methylation of the pyridine nitrogen of the vitamin nicotinamide.

AMINO ACID CONJUGATION

The major substrates for amino acid conjugation are benzoic acid and related aromatic carboxylic acids such as phenylacetic acid, phenoxyacetic acid, cinnamic acid, etc. (21). In humans, the major amino acid utilized in the conjugation is glycine; however, glutamine and taurine can also be cofactors. In birds, the major amino acid utilized is ornithine.

In most conjugations, it is the cofactor that is activated, but in amino acid conjugation it is the substrate that is activated, first by reacting with ATP to form an AMP conjugate, which is further converted to a CoA thio ester as shown in Figure 7.13.

As with glucuronidation, amino acid conjugation activity is very low in the newborn and that makes benzoic acid derivatives more toxic in them. For example, several years ago premature babies sometimes developed an often-fatal syndrome characterized by metabolic acidosis (22). This was eventually linked to the use of sterile water that contained a small amount of benzyl alcohol as a preservative for the administration of IV medication. Benzyl alcohol is readily metabolized to benzoic acid. In an adult, this is perfectly safe because the dose relative to weight is quite small and adults readily convert the benzoic acid to its glycine conjugate (commonly called hippuric acid) (Fig. 7.13). However, in the premature babies the dose relative to size was much greater and they were not able to conjugate the benzoic acid. Benzoic acid interferes with fatty acid β-oxidation and this is what led to the metabolic acidosis. With this understanding, a different diluent was used for administration of medication and the problem was eliminated.

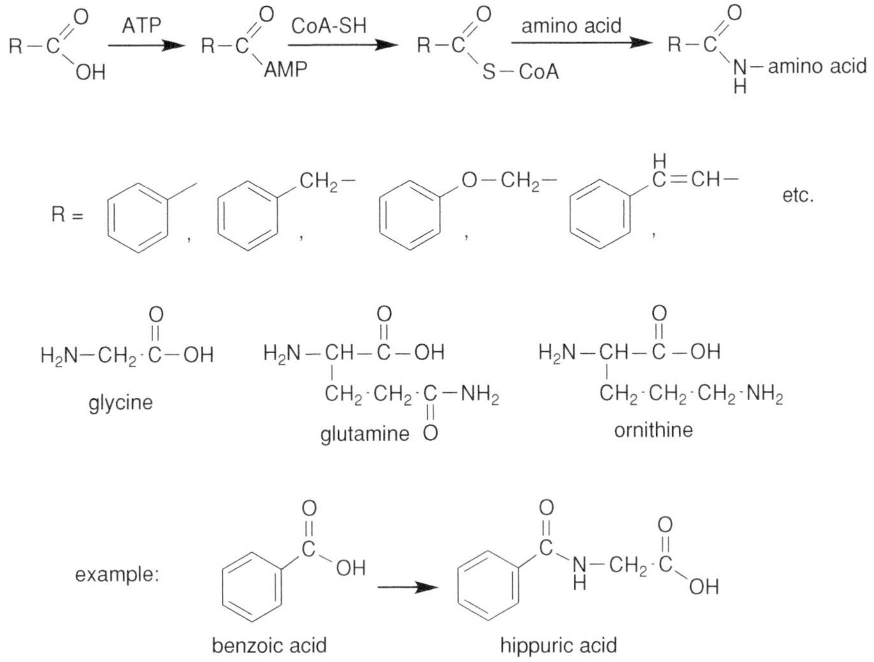

FIGURE 7.13 Reaction sequence leading to amino acid conjugation, the type of carboxylic acids that are substrates, and the type of amino acids that can conjugate. An example is the addition of glycine to benzoic acid to form hippuric acid.

CHIRAL INVERSION OF 2-ARYLPROPIONIC ACIDS

Many nonsteroidal anti-inflammatory drugs (NSAIDs) are substituted 2-arylpropionic acids. Most NSAIDs also have a chiral carbon next to the carboxylate and are administered as a racemic mixture of the two enantiomers. In general, the (S)-enantiomer is responsible for most of the antiinflammatory activity of these agents. It was found that the (R)-enantiomer is converted to the (S)-enantiomer but the reverse does not occur (23). As with amino acid conjugation, the pathway involves reaction with ATP to form an AMP ester, which is, in turn, converted to a Co-A ester, and it is the Co-A ester that undergoes chiral inversion (Fig. 7.14). Substrates include ibuprofen, naproxen, and fenoprofen.

Although this pathway involves the formation of an AMP adduct similar to the amino acid conjugation pathway, it is really not a conjugation pathway because the result is simply the inversion of configuration.

(R)-enantiomer   (S)-enantiomer

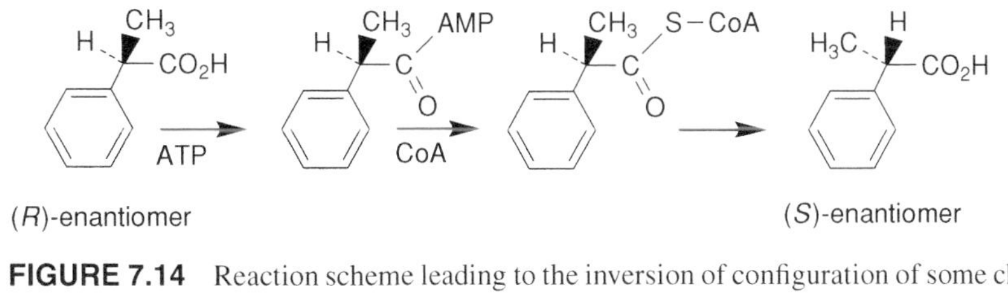

FIGURE 7.14 Reaction scheme leading to the inversion of configuration of some chiral carboxylic acids.

GLUTATHIONE CONJUGATION

In terms of a conjugating system, the glutathione S-transferases (GSTs) fulfill a unique role. They catalyze the reaction between the tripeptide glutathione and reactive electrophilic sites in molecules generated by metabolic processes or sites, such as α,β-unsaturated ketones or halogenated alkyl groups, in parent molecules that are susceptible to nucleophilic attack. Glutathione is a good soft nucleophile due to its cysteine thiol group, and it forms conjugates with drugs or, more commonly, drug metabolites that are electrophiles (Fig. 7.15). The reaction involves the anion of the sulfhydryl group, which is a weak acid with a pK_a of 9.1. Unlike other conjugation reactions, any drug that can form a glutathione conjugate in the presence of a glutathione transferase can also form a glutathione conjugate in the absence of a transferase. One of the roles of the transferase is to increase the fraction of glutathione thiol group that is ionized and this increases the rate of the reaction.

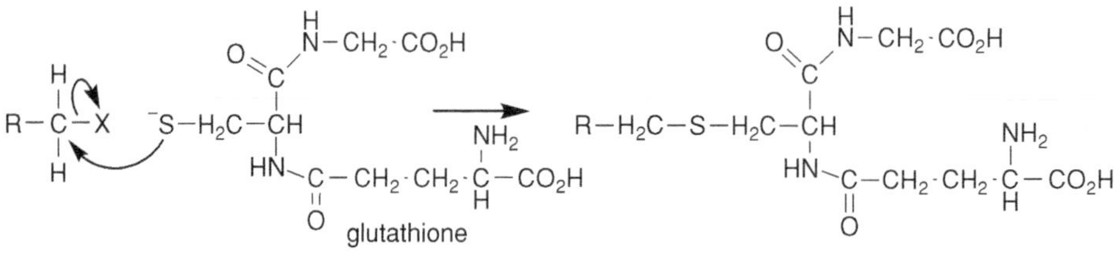

where X is a good leaving group such as a halogen, sulfate, etc.

FIGURE 7.15 Reaction scheme for the conjugation of a reactive electrophile (in this case with a good leaving group) with glutathione.

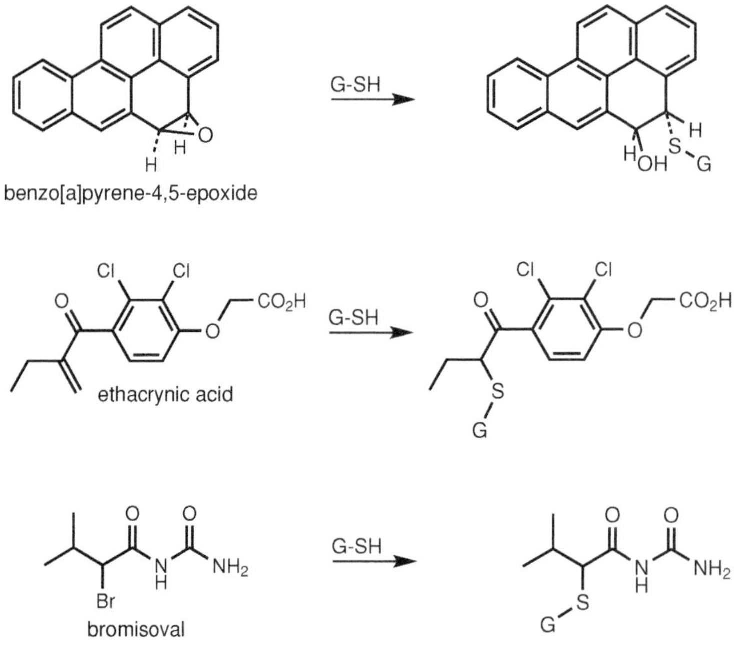

benzo[a]pyrene-4,5-epoxide

ethacrynic acid

bromisoval

FIGURE 7.16 Examples of electrophiles that form glutathione (GSH) conjugates.

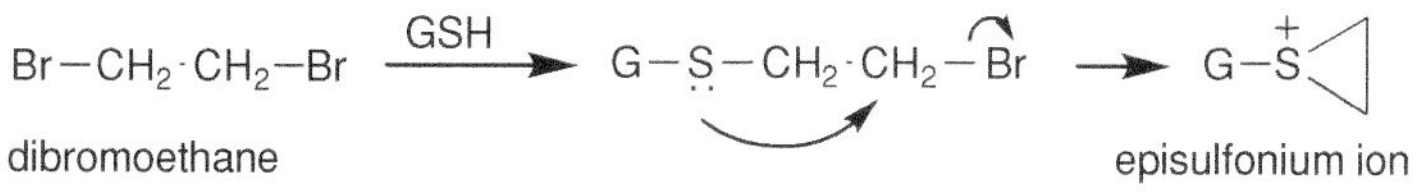

FIGURE 7.17 Conjugation of dibromoethane with glutathione (GSH) ultimately leads to a more reactive episulfonium ion.

Five classes of cytosolic mammalian GSTs, called alpha, mu, pi, theta, zeta and now designated as A, M, U, P, and Z, have been identified and categorized based on sequence identity (40% sequence identity within a class). At least three membrane-associated microsomal GSTs have also been isolated. Individual GSTs belonging to different classes have overlapping substrate specificities but metabolize common substrates at different rates. Another major difference is that different GSTs can be expressed in different tissues. For example, human GST1–1 is expressed in liver whereas human GSTM3–3 is not (24).

An example of three types of reactive substrates includes the metabolically generated epoxides of polycyclic aromatic hydrocarbons, e.g., benzo[*a*]pyrene-4,5-epoxide (25), the diuretic ethacrynic acid, and the hypnotic agent bromisoval as shown in Figure 7.16.

It is more difficult to describe the range of drugs and metabolites that are substrates for this conjugation reaction because it is based on the substrate being an electrophile rather than a specific functional group. What makes a drug or metabolite an electrophile will be discussed further in Chapter 8. In most cases, conjugation with glutathione decreases the toxicity of an agent; however, there are cases such as dibromoethane in which glutathione conjugation leads to a more reactive species, in this case an episulfonium ion (Fig. 7.17).

The primary mission of the GSTs appears to be to defend the organism from toxicities resulting from the covalent modification of critical bio-macromolecules through catalyzing the reaction of glutathione with chemically reactive foreign species. The cofactor glutathione also plays another major defensive role in protecting the organism from destructive effects of lipid peroxidation. In addition to its properties as a potent nucleophile, glutathione also has a significant reductive capacity by virtue of the thiol group. It is this reductive capacity that makes glutathione an efficient scavenger of reactive one electron species such as reactive oxygen species and radicals generated by lipid peroxidation.

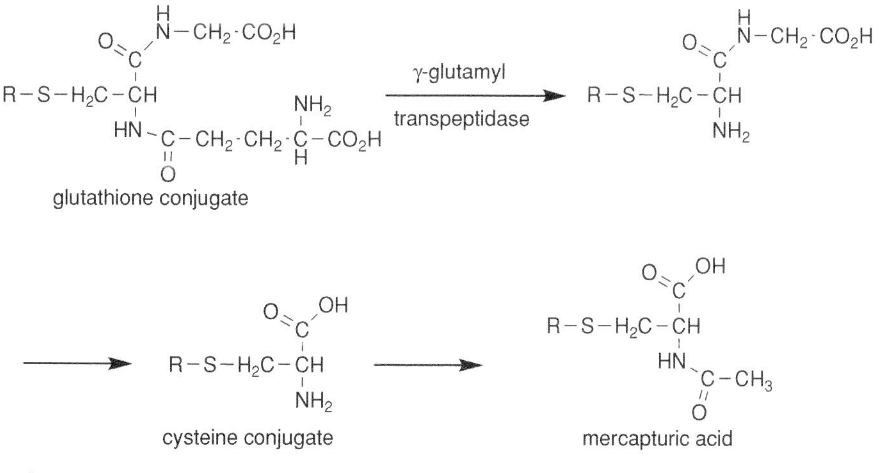

FIGURE 7.18 Reaction scheme by which glutathione conjugates are converted to mercapturic acid conjugates.

CONVERSION OF GLUTATHIONE CONJUGATES TO MERCAPTURIC ACIDS

Glutathione conjugates are often converted to mercapturic acids before excretion in urine. Mercapturic acids are simply N-acetylcysteine conjugates. This process involves hydrolytic removal of glutamic acid and glycine leaving the cysteine conjugate, and then the cysteine conjugate is N-acetylated in the kidney before excretion (Fig. 7.18). Although this is the classic pathway for glutathione conjugates, the urine usually contains a mixture of parent glutathione conjugate, cysteine conjugate, and mercapturic acids. In addition, many glutathione conjugates, especially if they are of high molecular mass, are excreted into bile. This also is not strictly a conjugation pathway but rather the further metabolism of a conjugate.

CYSTEINE CONJUGATE β-LYASE

Cysteine conjugates can also be degraded by the enzyme cysteine conjugate β-lyase, which involves pyridoxal as a cofactor as shown in Figure 7.19 (26).

In many cases, the product is toxic as in the case of the conjugate of trichloroethylene, which is thought to be responsible for the aplastic anemia induced in calves fed trichloroethylene-extracted soybean oil meal (27). The unstable product shown in brackets (Fig. 7.20) has the potential to lose HCl to form a reactive thioketene or tautomerize to form the reactive chlorothioacetyl chloride (27).

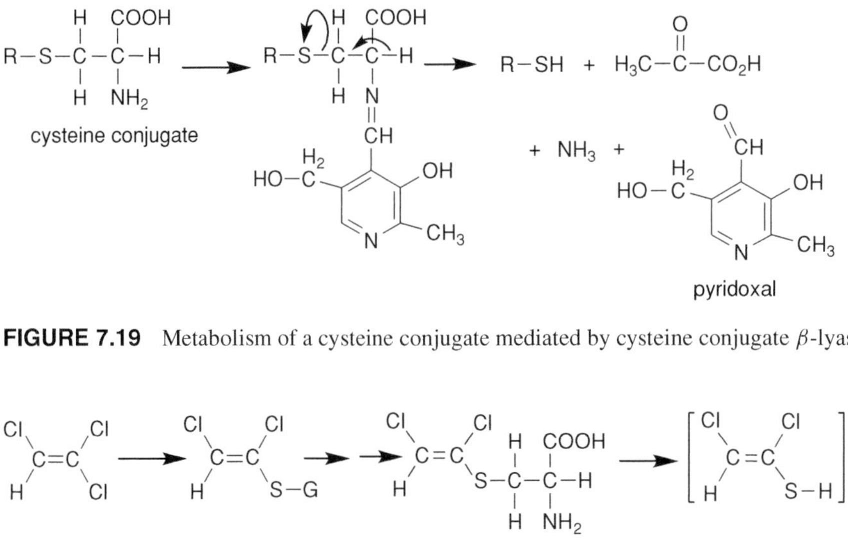

FIGURE 7.19   Metabolism of a cysteine conjugate mediated by cysteine conjugate β-lyase.

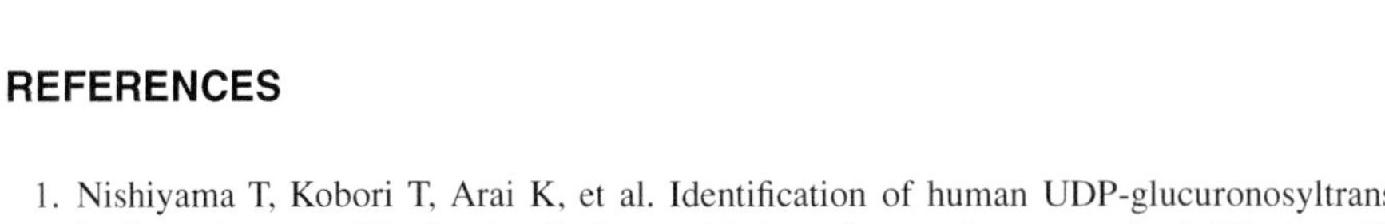

FIGURE 7.20   Metabolism of trichloroethylene to a toxic product. The last step in this sequence is mediated by cysteine conjugate β-lyase.

REFERENCES

1. Nishiyama T, Kobori T, Arai K, et al. Identification of human UDP-glucuronosyltransferase isoform(s) responsible for the C-glucuronidation of phenylbutazone. Arch Biochem Biophys 2006;454(1):72–79.
2. Tukey RH, Strassburg CP. Human UDP-glucuronosyltransferases: metabolism, expression, and disease. Annu Rev Pharmacol Toxicol 2000;40:581–616.

3. Chiu SH, Huskey SW. Species differences in N-glucuronidation. Drug Metab Dispos 1998;26(9):838–847.

4. Aarbakke J. Clinical pharmacokinetics of phenylbutazone. Clin Pharmacokinet 1978;3(5):369–380.

5. Tephly TR, Green MD. UDP-Glucuronosyltransferases. In: Levy RH, Thummel KE, Trager WF, et al., eds. Metabolic Drug Interactions. Philadelphia, PA: Lippincott, Williams & Wilkins; 2000.

6. Johnson TN. The development of drug metabolising enzymes and their influence on the susceptibility to adverse drug reactions in children. Toxicology 2003;192(1):37–48.

7. Andersen G, Christrup L, Sjogren P. Relationships among morphine metabolism, pain and side effects during long-term treatment: an update. J Pain Symptom Manage 2003;25(1):74–91.

8. Weinshilboum RM, Raftogianis RB. Sulfotransferases and methyltransferases. In: Levy RH, Thummel KE, Trager WF, et al., eds. Metabolic Drug Interactions. Philadelphia, PA: Lippincott, Williams & Wilkins; 2000.

9. Goldstein DS, Eisenhofer G, Kopin IJ. Sources and significance of plasma levels of catechols and their metabolites in humans. J Pharmacol Exp Ther 2003;305(3):800–811.

10. Baker CA, Uno H, Johnson GA. Minoxidil sulfation in the hair follicle. Skin Pharmacol 1994;7(6):335–339.

11. Weber WW, Hein DW. N-acetylation pharmacogenetics. Pharmacol Rev 1985;37(1):25–79.

12. Hanna PE. N-acetyltransferases, O-acetyltransferases, and N,O-acetyltransferases: enzymology and bioactivation. Adv Pharmacol 1994;27:401–430.

13. Hein DW. Molecular genetics and function of NAT1 and NAT2: role in aromatic amine metabolism and carcinogenesis. Mutat Res 2002;506–507:65–77.

14. Boukouvala S, Fakis G. Arylamine N-acetyltransferases: what we learn from genes and genomes. Drug Metab Rev 2005;37(3):511–564.

15. Grant DM, Blum M, Beer M, et al. Monomorphic and polymorphic human arylamine N-acetyltransferases: a comparison of liver isozymes and expressed products of two cloned genes. Mol Pharmacol 1991;39(2):184–191.

16. Zhang N, Liu L, Liu F, et al. NMR-based model reveals the structural determinants of mammalian arylamine N-acetyltransferase substrate specificity. J Mol Biol 2006;363(1):188–200.

17. Windmill KF, Gaedigk A, Hall PM, et al. Localization of N-acetyltransferases NAT1 and NAT2 in human tissues. Toxicol Sci 2000;54(1):19–29.

18. Carreon T, Ruder AM, Schulte PA, et al. NAT2 slow acetylation and bladder cancer in workers exposed to benzidine. Int J Cancer 2006;118(1):161–168.

19. Weinshilboum RM, Otterness DM, Szumlanski CL. Methylation pharmacogenetics: catechol O-methyltransferase, thiopurine methyltransferase, and histamine N-methyltransferase. Annu Rev Pharmacol Toxicol 1999;39:19–52.

20. Crooks PA, Godin CS, Damani LA, et al. Formation of quaternary amines by N-methylation of azaheterocycles with homogeneous amine N-methyltransferases. Biochem Pharmacol 1988;37(9):1673–1677.

21. Testa B, Jenner P. Drug Metabolism: Chemical and Biochemical Aspects. New York, NY: Dekker; 1976.

22. Menon PA, Thach BT, Smith CH, et al. Benzyl alcohol toxicity in a neonatal intensive care unit. Incidence, symptomatology, and mortality. Am J Perinatol 1984;1(4):288–292.

23. Wsol V, Skalova L, Szotakova B. Chiral inversion of drugs: coincidence or principle? Curr Drug Metab 2004;5(6):517–533.

24. Eaton DL, Bammler TK. Glutathione S-transferases. In: Levy RH, Thummel KE, Trager WF, et al., eds. Metabolic Drug Interactions. Philadelphia, PA: Lippincott, Williams & Wilkins; 2000.

25. Foureman GL, Hernandez O, Bhatia A, et al. The stereoselectivity of four hepatic glutathione S-transferases purified from a marine elasmobranch (Raja erinacea) with several K-region polycyclic arene oxide substrates. Biochim Biophys Acta 1987;914(2):127–135.

26. Anders MW. Glutathione-dependent bioactivation of haloalkanes and haloalkenes. Drug Metab Rev 2004;36(3–4):583–594.

27. Anderson PM, Schultze MO. Cleavage of S-(1,2-dichlorovinyl)-L-cysteine by an enzyme of bovine origin. Arch Biochem Biophys 1965;111(3):593–602.

8

Reactive Metabolites

Although it is essential for an organism to have enzymes that can convert lipophilic molecules into a form that can be eliminated as well as to convert toxic molecules to less toxic molecules, the process is not perfect and virtually any metabolic enzyme is also capable of converting some molecules to chemically reactive species that can be toxic. Chemically reactive metabolites can react with DNA, proteins, or other molecules leading to mutations, cancer, birth defects, and a variety of other types of toxicity. Therefore, it is important to understand the pathways that can lead to reactive metabolites and to be able to predict when a specific metabolite will be chemically reactive. In particular, many of the adverse reactions, especially idiosyncratic reactions caused by drugs, are believed to be due to reactive metabolites. For this reason, a major aspect of drug metabolism studies during drug development involves prediction of the potential that a drug candidate will form a reactive metabolite as well as screening drug candidates to determine if reactive metabolites are actually formed.

Most reactive metabolites are electrophiles or free radicals. An electrophile is a molecule that is electron deficient and reacts with nucleophiles, which usually have a negative charge or a lone pair of electrons that can form a bond to the electrophile. Although there may be cases in which reactive metabolites are strong nucleophiles rather than electrophiles, there are no clear examples.

It is sometimes useful to classify electrophiles as being hard or soft depending on how concentrated or diffuse the electron deficiency is. The utility of this concept arises from the observation that there is some selectivity of soft electrophiles to react with soft nucleophiles, whereas hard electrophiles demonstrate some selectivity to react with hard nucleophiles. The major biological soft nucleophiles are thiols such as glutathione, which are softer than nitrogen nucleophiles because sulfur is larger than nitrogen and, therefore, its electron cloud is more diffuse. For example, Michael acceptors such as acrolein and the imidoquinone of acetaminophen (discussed later in this chapter) are soft electrophiles and react almost exclusively with the soft nucleophile, glutathione, and other endogenous sulfhydryl-containing molecules. In contrast, trifluoroacetyl chloride, the reactive metabolite of halothane, reacts readily with amino-containing nucleophiles such as the amino group of lysine (see later in Fig. 8.7). Although this is a useful concept, it is sometimes difficult to apply with precision in practice. Very reactive metabolites show little selectivity and some are so reactive that they react almost exclusively with the enzyme that formed them. This often leads to inactivation of the enzyme, and this is referred to as mechanism-based or suicide inhibition.

There are many factors that can result in a reactive metabolite and modify its reactivity including the presence of a good leaving group, ring strain, a double bond conjugated with a carbonyl group (Michael acceptors), and the presence of electron-withdrawing groups; the following is a more detailed description of these concepts.

GOOD LEAVING GROUPS

In the reaction between a reactive electrophile and a nucleophile, a new bond is formed and this usually requires a bond to be broken so that the proper valence can be maintained on the atoms involved in the reaction. If the electrophile has a good leaving group at the electrophilic center, this facilitates the reaction and makes the electrophile more reactive. The leaving group usually leaves with a negative charge and, therefore, the ability of the group to stabilize a negative charge is a major factor making it a good leaving group. In general, good leaving groups are strong acids when protonated because what makes them a strong acid is their ability to accept a negative charge on loss of the acidic proton. For example, chloride and sulfonate are good leaving groups and hydrochloric acid and sulfonic acid are strong acids. Sulfuric acid (H_2SO_4) is a very strong acid but bisulfate (HSO_4^-), while still strong, is a weaker acid because it acquires a second negative charge on loss of the proton; therefore, sulfate is not as good a leaving group as chloride.

The following are examples of drugs or metabolites that are reactive because of the presence of good leaving groups. Busulfan is a bifunctional alkylating drug used to treat chronic myelogenous leukemia (Fig. 8.1). It reacts with DNA, and being bifunctional it can lead to DNA cross-linking; it also reacts with glutathione (1). N-acetylaminofluorene was developed as an insecticide until it was found to be a carcinogen. It was discovered that this is due to bioactivation leading to a reactive metabolite that binds to DNA. The pathway involves oxidation of the amide nitrogen to form a hydroxamic acid as mentioned in Chapter 4; this is further activated by conjugation with sulfate as shown in Figure 8.1 (2). Likewise the carcinogen safrole, a component of sassafras, was found to be bioactivated through hydroxylation of the benzylic carbon followed by sulfation (3).

Not only is the leaving group important, but if the reaction has S_N1 character, the stability of the positive charge left behind is also important. In the examples above, a primary carbocation is difficult to form and therefore the reaction of busulfan with glutathione would likely be a S_N2-type reaction, whereas the nitrenium ion and carbocation formed from N-acetylaminofluorene and safrole, respectively, are relatively stable and likely to be S_N1-type reactions.

This principle is also seen in comparing the reactivity of ethyl chloride, vinyl chloride, allyl chloride, chlorobenzene, benzyl chloride, and 2,4-dinitrochlorobenzene whose structures are shown in Figure 8.2. Ethyl chloride is not very reactive because chloride is not quite as good a leaving group as the sulfonate in the example of busulfan. Vinyl chloride is even less reactive because the sp^2-hybridized carbon has more s character making the C–Cl bond stronger and vinyl carbocations more difficult to form. In fact, the C–Cl bond is so unreactive that addition reactions across the double bond are the primary mode of reaction for the molecule and not substitution of the C–Cl bond. In contrast, allyl chloride is reactive because the carbocation has resonance stabilization analogous to that of safrole.

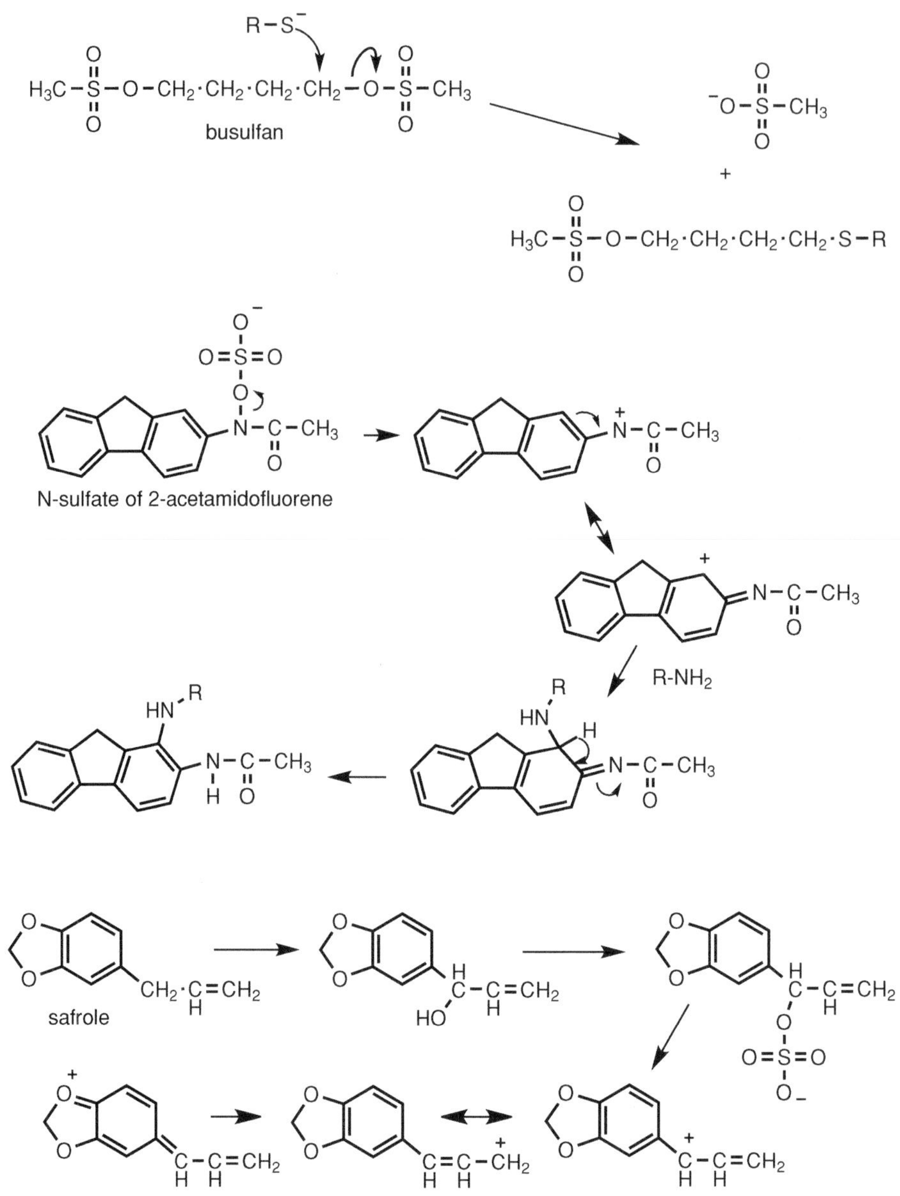

FIGURE 8.1 Examples of reactive metabolites that involve sulfate or sulfonate as the leaving group.

Chlorobenzene is analogous to vinyl chloride and does not undergo substitution reactions, e.g., by reacting directly with glutathione, whereas benzyl chloride is analogous to allyl chloride and is reactive. Although 2,4-dinitrochlorobenzene could be viewed as analogous to vinyl chloride, the two strongly electron-withdrawing nitro groups make the molecule very electrophilic and this molecule reacts quite readily with glutathione in a more S_N2 manner, i.e., there is no carbocation intermediate.

It is also useful to compare the bioactivation of aminobiphenyl and sulfamethoxazole. Aminobiphenyl is a carcinogen found in cigarette smoke. Its bioactivation is similar to that of N-acetylaminofluorene in which the first step is oxidation to form a hydroxylamine. The

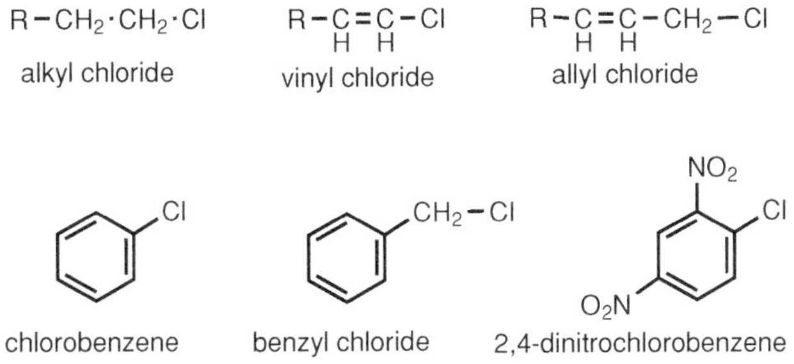

FIGURE 8.2   Structures of different electrophiles with the same chloride leaving group.

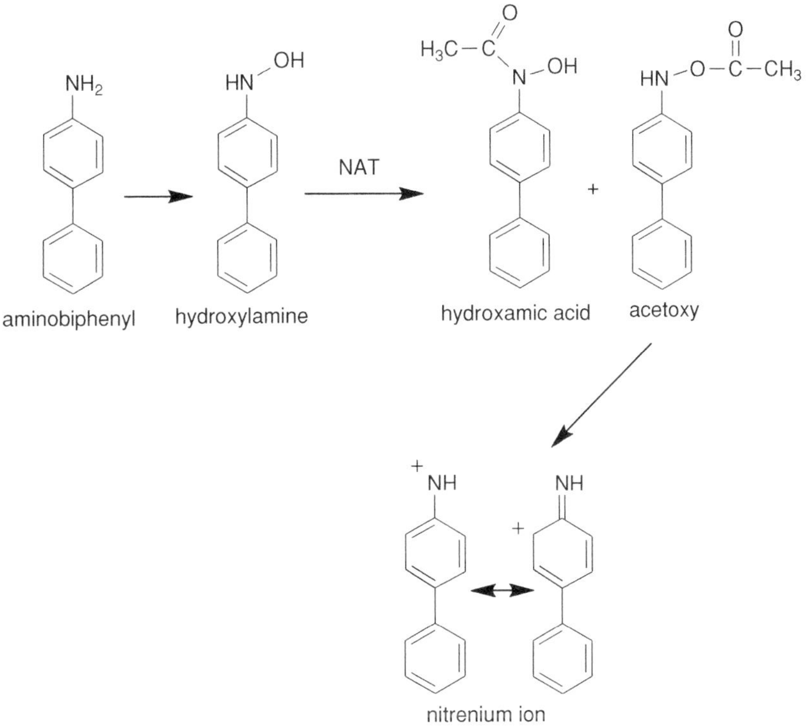

FIGURE 8.3   Metabolism of aminobiphenyl to a reactive nitrenium ion.

hydroxylamine is not sufficiently reactive, but it can be *o*-acetylated and loss of acetate leads to a relatively stable but reactive nitrenium ion as shown in Figure 8.3.

Sulfamethoxazole is an antibiotic that also undergoes N-oxidation followed by acetylation; however, the electron-withdrawing effect of the para sulfonamide group (three highly electronegative atoms attached to sulfur) makes it too difficult to form a nitrenium ion (Fig. 8.4). The major reactive metabolite of sulfamethoxazole appears to be the nitroso metabolite (4), which can react with glutathione to form a sulfinamide as shown in Figure 8.4.

Almost all drugs that contain a primary aromatic amine or aromatic nitro group are associated with a significant incidence of adverse reactions (5), presumably because they are oxidized or reduced, respectively, to similar reactive intermediates; examples are shown in Figure 8.5.

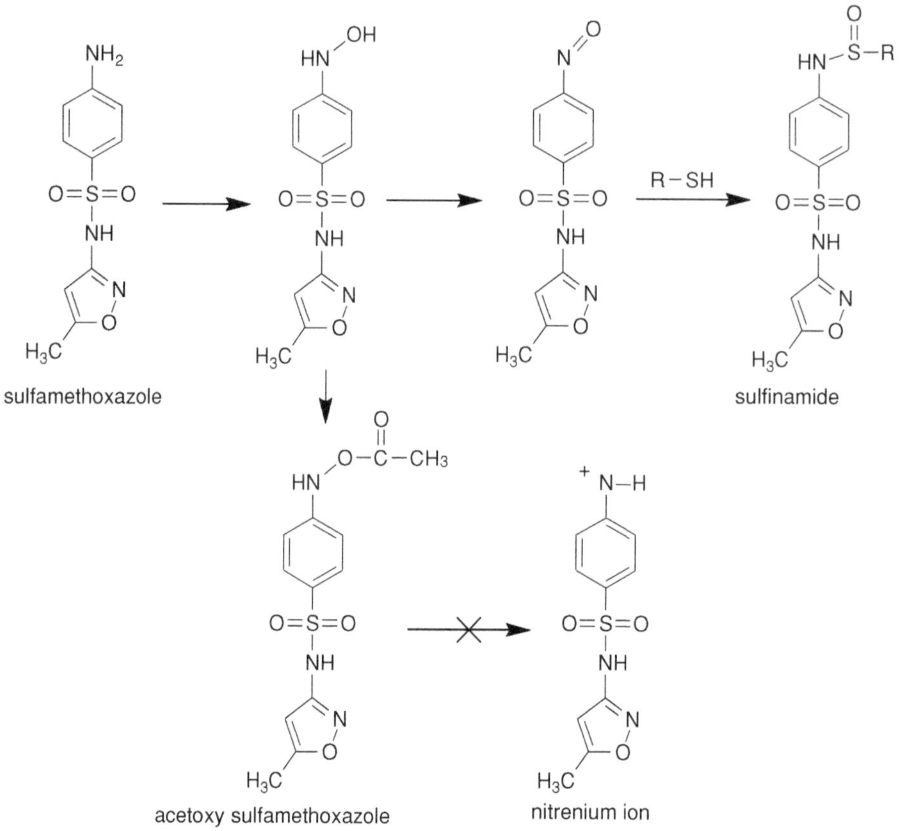

FIGURE 8.4 Sulfamethoxazole cannot readily form a nitrenium ion and the major reactive metabolite is the nitroso metabolite.

In contrast to sulfamethoxazole, the nitrenium ion formed by oxidation of clozapine (Fig. 8.6) is very delocalized and very stable with a half-life of almost a minute in buffer (6). In fact, one could argue whether it should be called a nitrenium ion because most of the charge density is not on the nitrogen as drawn. It is spread throughout the aromatic system as demonstrated by the two major glutathione adducts that are formed as shown in Figure 8.6. This can be considered an S_N1-type reaction because the half-life of the nitrenium ion is greater than that of the N-chloro precursor. The positive charge is also delocalized on the three nitrogens, and although glutathione likely also reacts with one or more of the nitrogens, the product would not be stable and would react with another molecule of glutathione to regenerate clozapine and form oxidized glutathione.

In contrast, the reactivity of trifluoroacetyl chloride, the reactive metabolite of halothane discussed in Chapter 4 under oxidative dehalogenation (Fig. 8.7), is due to the electron-withdrawing effect of the carbonyl and trifluoromethyl groups, which makes it very electrophilic, more reactive than most other molecules that have chloride as the leaving group (Fig. 8.7).

Nitrogen gas is such a good leaving group that it can lead to the formation of electrophiles that are quite difficult to form such as a primary carbocation as mentioned in Chapter 4 (Fig. 4.87) and also as illustrated in Figure 8.8. Oxidation of aminobenzotriazole produces two molecules of nitrogen gas and benzyne, which is an alkyne that is prevented from being linear by being part of a ring system (Fig. 8.8). This makes it very reactive and

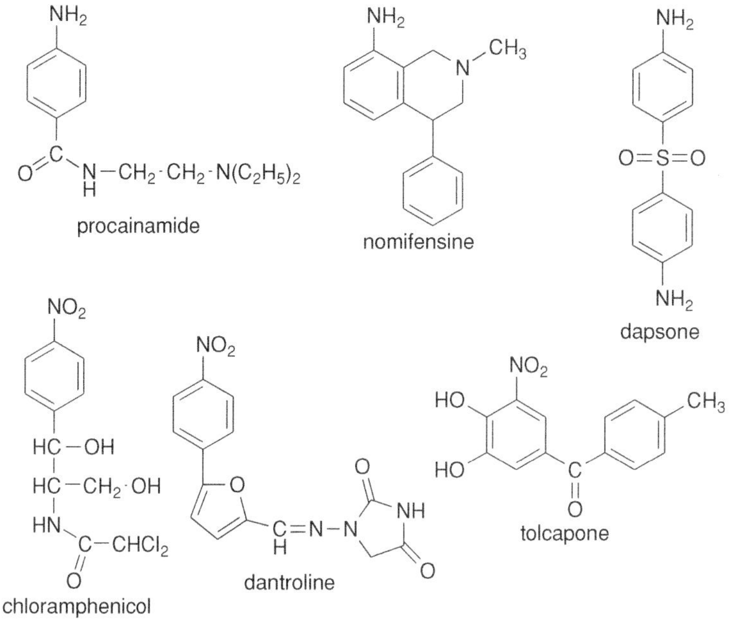

FIGURE 8.5 Examples of drugs that have aromatic amine or nitro functional groups and are associated with a significant incidence of adverse drug reactions.

it inactivates the cytochrome P450 that formed it. Aminobenzotriazole and its derivatives are one of the most effective general P450 inhibitors (7). Another good leaving group, presumably SO_2, is formed by the oxidation of thiono sulfur compounds such as the thiourea drug [N-(5-chloro-2-methylphenyl)-N-(2-methylpropyl)thiourea] as shown in Figure 8.8. In this figure, the number of oxygens is designated by x because it has not been proven but it is most likely two (8).

    Although not a very good leaving group, glucuronic acid is a better leaving group than hydroxide, and acyl glucuronides react slowly with protein amino groups as shown in Figure 8.9 (9). In addition, acyl glucuronides can rearrange, i.e., the glucuronyl group can migrate to adjacent ring hydroxyl groups, exposing a free aldehyde group that can react reversibly with protein amino groups. This product can undergo an Amadori rearrangement, which "locks" in the bond formed between the glucuronide and the protein nucleophile (Fig. 8.9). Unlike the direct reaction of an acyl glucuronide with a nucleophile, this reaction also results in the glucuronic acid moiety bound to the nucleophile. Furthermore, carboxylic acids can form Co-A thioesters, which can also react with protein amino groups (10). It has been suggested that one or more of these reactions is responsible for the relatively high incidence of idiosyncratic adverse reactions associated with some drugs that are carboxylic acids; however, there is little direct evidence to support this hypothesis, and in most cases alternative reactive metabolites are known.

RING STRAIN

Ring strain can also increase the reactivity of a compound. The normal bond angle of an sp^3-hybridized carbon is 109°; therefore, a carbon in a three-membered ring in which the bond angle is forced to be 60° is under a considerable amount of strain and a reaction that opens the ring is facilitated. An example of such increased reactivity is the

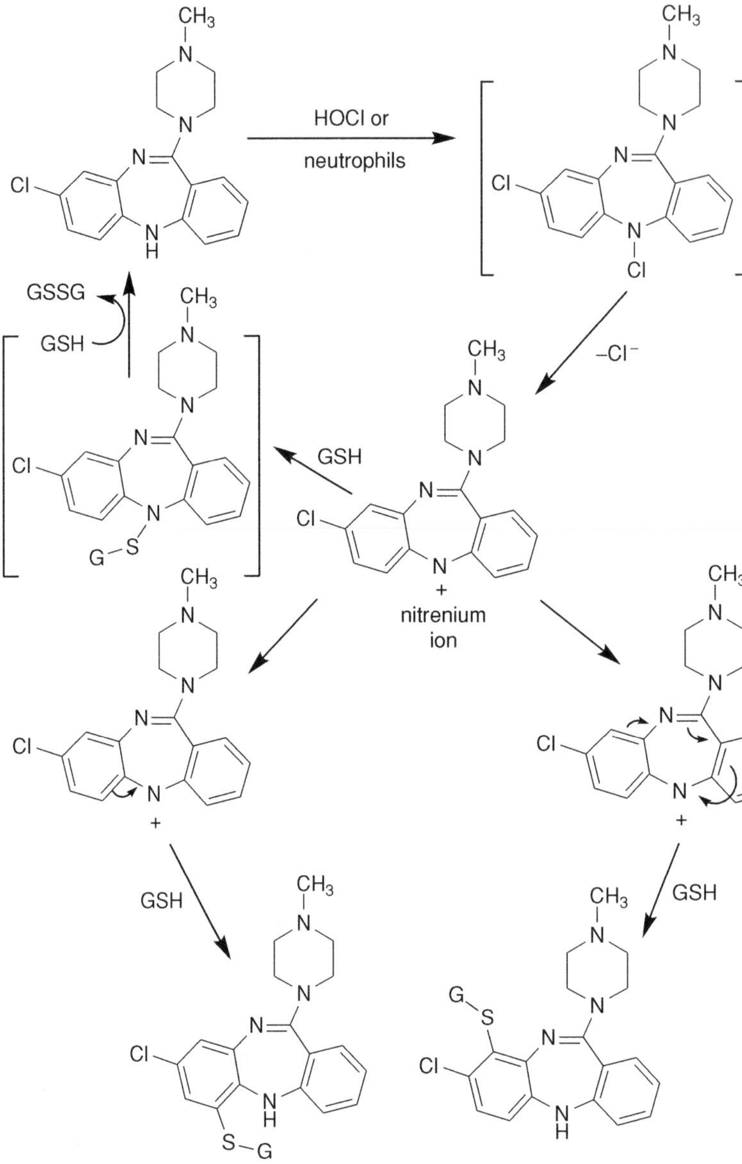

FIGURE 8.6 Oxidation of clozapine by activated neutrophils to a relatively stable nitrenium ion and subsequent reaction with glutathione.

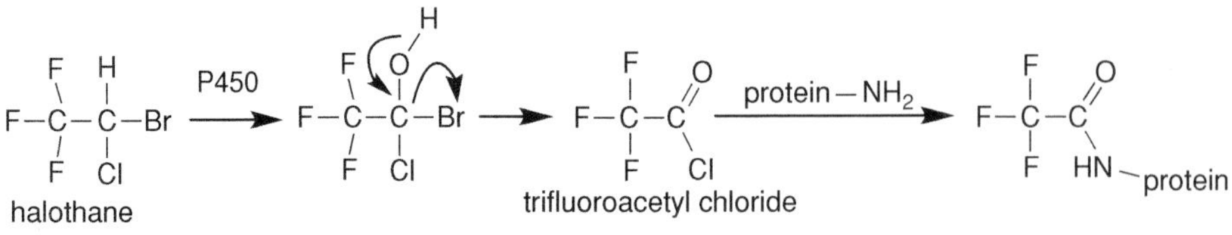

FIGURE 8.7 Oxidation of halothane to the highly reactive trifluoroacetyl chloride.

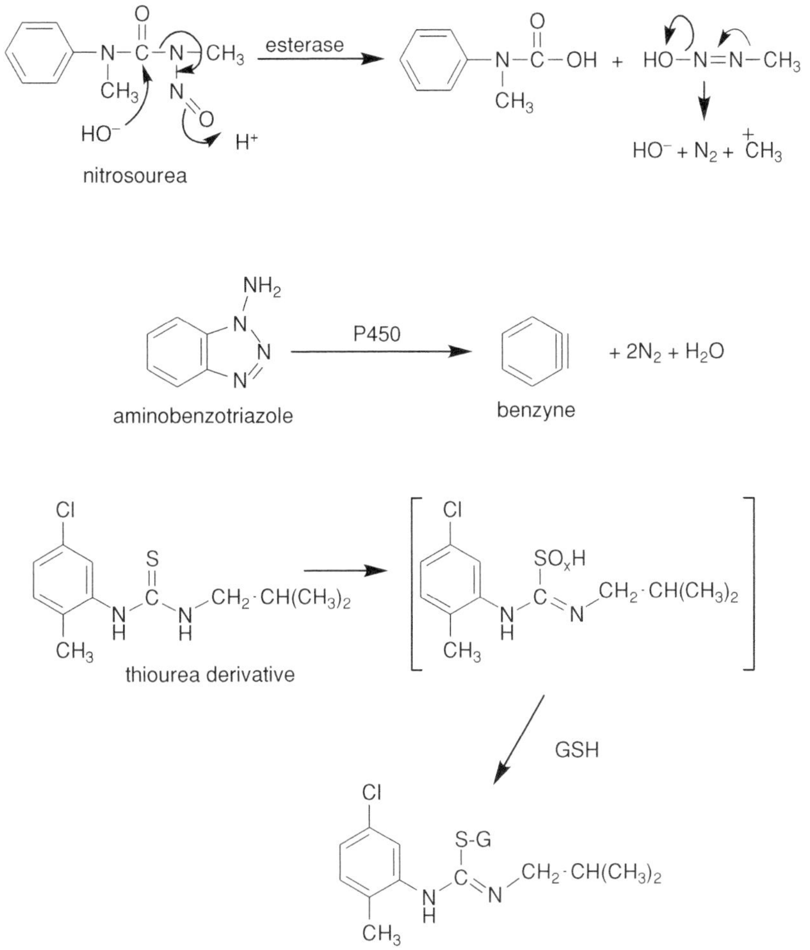

FIGURE 8.8 Examples of compounds that form reactive metabolites because of the loss of nitrogen gas or SO$_2$ which allows the formation of species that would otherwise be difficult to form.

nitrogen mustard, mechlorethamine, which is an alkylating agent used to treat cancer. An intramolecular reaction leads to an aziridinium ion which is both positively charged and also has ring strain that is relieved when it reacts with a nucleophile as shown in Figure 8.10. Conjugation with glutathione usually inactivates a reactive compound. However, in the case of 1,2-dibromoethane, reaction of glutathione produces a sulfur mustard analogous to mechlorethamine which undergoes an intramolecular reaction to produce a reactive episulfonium ion (11) as mentioned in Chapter 7.

Unlike aziridinium ions and episulfonium ions, epoxides do not have a positive charge and are less reactive. However, epoxides formed from the oxidation of aromatic rings (often referred to as arene oxides) are more reactive than most epoxides because conjugation leads to stabilization of the intermediate, and the reaction can lead to rearomatization of the ring. Even so, the epoxide formed by the oxidation of benzene has a surprisingly long half-life in blood and can reach sites distant from where it is formed (12). Epoxides are often detoxified by epoxide hydrolase, but "bay region " diol epoxides of polycyclic aromatic hydrocarbons are more toxic (Fig. 8.11), in part, because they are protected from inactivation by epoxide hydrolase owing to steric hindrance (13). The β-lactam ring of penicillins and

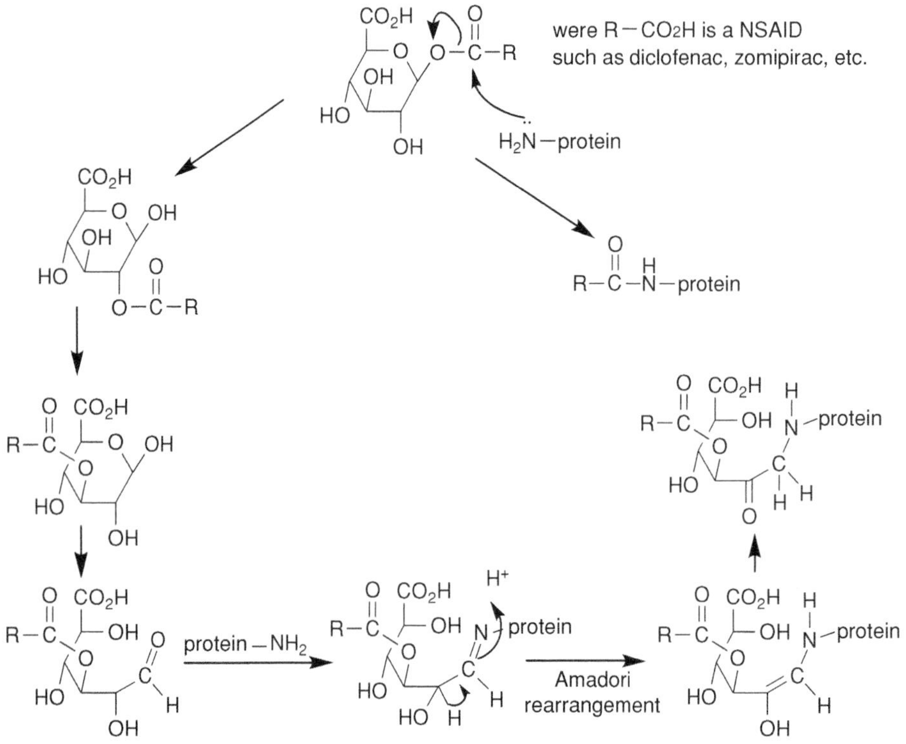

FIGURE 8.9 Acyl glucuronides can act as electrophiles both through a direct S_N2 reaction and after rearrangement of the acyl glucuronide.

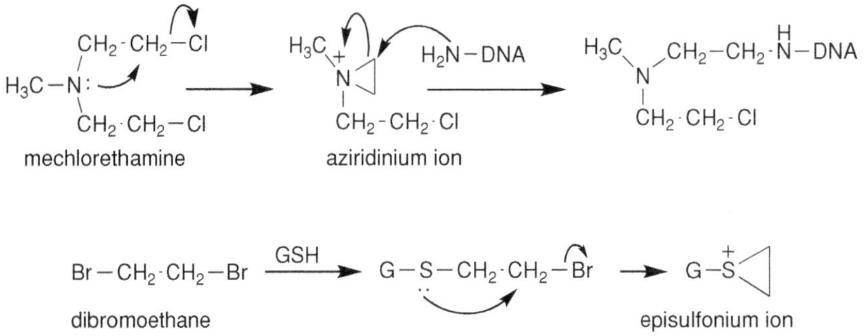

FIGURE 8.10 Examples of reactive metabolites that involve ring strain.

cephalosporins are reactive with amino (Fig. 8.11) and sulfhydryl groups because of ring strain, and this is responsible for their association with allergic reactions; however, their reactions with proteins are quite slow (14).

MICHAEL ACCEPTORS

There does not have to be a leaving group if the nucleophile reacts with a double bond; however, in general, alkenes are not very reactive toward nucleophiles. An exception is if the double bond is polarized by conjugation with a carbonyl group. The reaction of

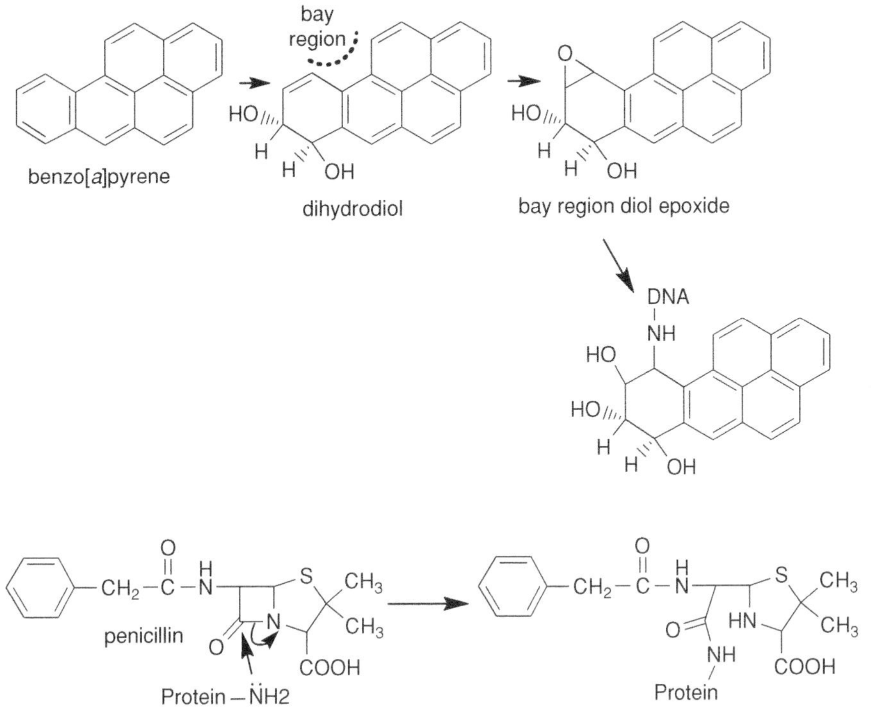

FIGURE 8.11   Penicillin and the reactive metabolite of benzo[*a*]pyrene are also reactive because of ring strain.

a nucleophile with a C=C that is conjugated with a carbonyl group is called a Michael addition. The simplest example is acrolein as shown in Figure 8.12. Felbamate, an anticonvulsant associated with aplastic anemia and liver toxicity, is metabolized to phenylacrolein (Fig. 8.12), and this metabolite is presumably responsible for the adverse reactions associated with this drug (15). Another example is terbinafine, which undergoes N-dealkylation to a Michael acceptor with extended conjugation (Fig. 8.12). In this case, glutathione can add 1,6 (the carbon that is attacked is 6 carbons from the carbonyl oxygen) and the product is still a Michael acceptor (16). This reactive glutathione conjugate is likely concentrated in bile where it may be responsible for the cholestatic hepatotoxicity that is associated with this drug. Finally, abacavir appears to undergo a combination of alcohol oxidation and double bond shift mediated by alcohol dehydrogenase to produce an acrolein analog, which of course is a reactive Michael acceptor (17).

Oxidation of a furan ring leads to a Michael acceptor as shown in Figure 8.13 (18). One example of a drug that contains a furan ring is furosemide. It causes hepatotoxicity in rodents but is relatively safe at normal doses in humans.

The reactions of nucleophiles with benzoquinone and related compounds can also be viewed as Michael reactions. Benzoquinone is one of the reactive metabolites of benzene, a solvent also associated with aplastic anemia (Fig. 8.14). A similar reactive metabolite is responsible for the hepatotoxicity of acetaminophen (Fig. 4.71), the most common cause of acute liver failure; however, most of this reactive metabolite is detoxified by reaction with glutathione, and it is only when glutathione is depleted to approximately 10% of the normal level that significant toxicity ensues.

If an even broader definition of a Michael reaction is used, iminoquinones (19), quinone iminium ions (20), quinone methides (21), etc. can also be viewed as Michael

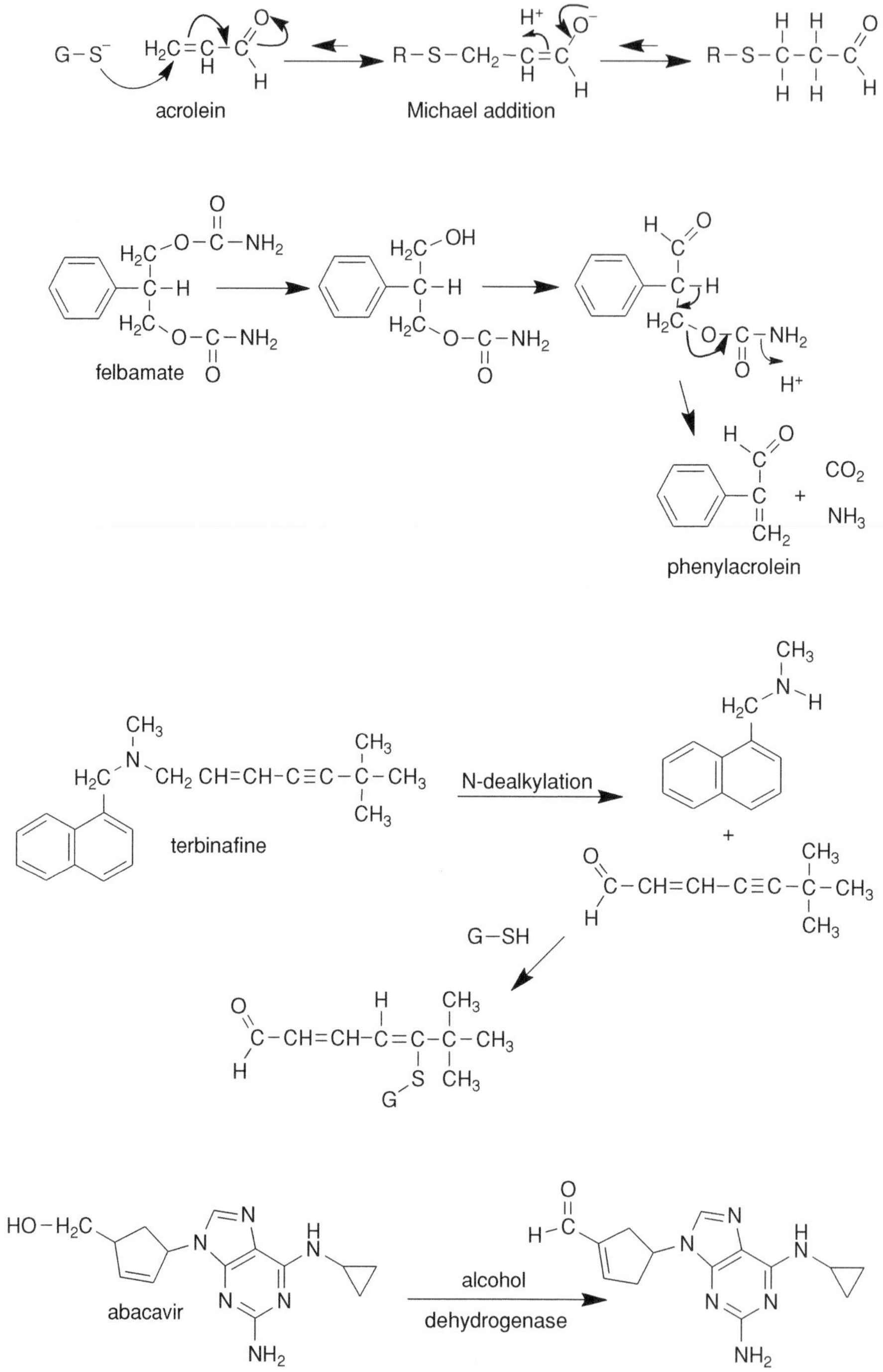

FIGURE 8.12 Michael reaction of glutathione with acrolein and acrolein-like reactive metabolites of felbamate, terbinafine, as well as the formation of a Michael acceptor metabolite of abacavir.

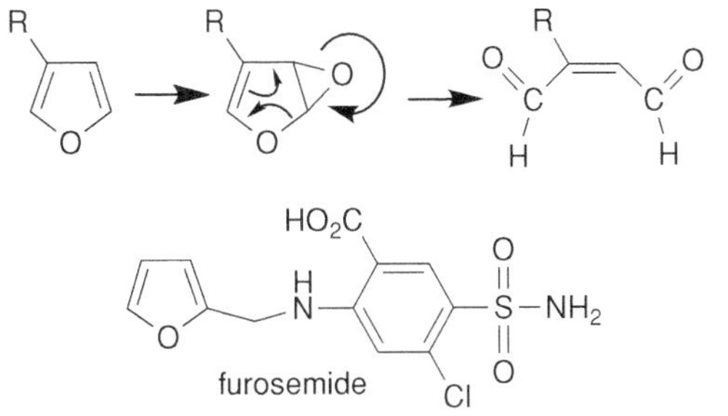

FIGURE 8.13 Oxidation of furan-containing drugs can lead to a reactive Michael acceptor.

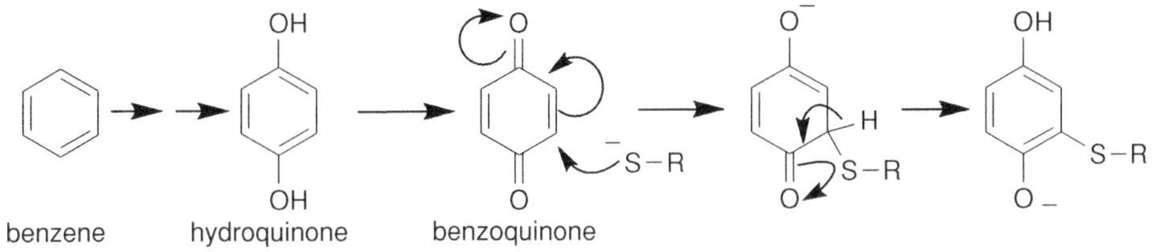

FIGURE 8.14 Oxidation of benzene forms a reactive Michael acceptor.

acceptors; examples are shown in Figure 8.15. This type of reactive metabolite is very common because many drugs have aromatic rings with an oxygen-, nitrogen-, or methylene group–containing substituent, and oxidation of such molecules usually leads to ortho or para oxidation; further oxidation leads to a quinone or quinone analog.

Even the S-oxide of a thiophene can be viewed as a Michael acceptor as illustrated in Figure 8.16 (22). An interesting example is the reactive metabolite of zileuton, a drug associated with liver toxicity, which is formed by N-dealkylation followed by S-oxidation as shown in Figure 8.16 (23). An S-oxide was proposed as the reactive metabolite of tienilic acid, a drug that was withdrawn from the market because of hepatotoxicity; however, more recent data point to an epoxide of the thiophene as being responsible for this idiosyncratic reaction (24). Another thiophene is ticlopidine, which is associated with agranulocytosis and aplastic anemia. It appears that a reactive S-chloro metabolite formed by neutrophils and analogous to the S-oxide may be responsible for this toxicity (25).

A related type of reactive metabolite is formed from 3-methylindoles. Microorganisms in the stomach of ruminants can convert L-tryptophan into 3-methylindole, which can cause pulmonary edema and death in cattle (26). The reactive metabolite is shown in Figure 8.17. This basic structure also occurs in some drugs such as zafirlukast, and this is presumably responsible for the idiosyncratic adverse reactions associated with this drug (27).

ISOCYANATES AND ISOTHIOCYANATES

Another type of activated double bond is found in isocyanates and isothiocyanates. Methylisocyanate was responsible for a disaster in Bhopal, India, in 1984 when approximately 40 tons of the gas were accidentally released leading to the death of thousands of people and injury to many more. Some have called this the worst industrial accident in history. Methylisocyanate reacts with nucleophiles as shown in Figure 8.18—in this case with a thiol. This reaction is reversible, and therefore a glutathione conjugate can act to

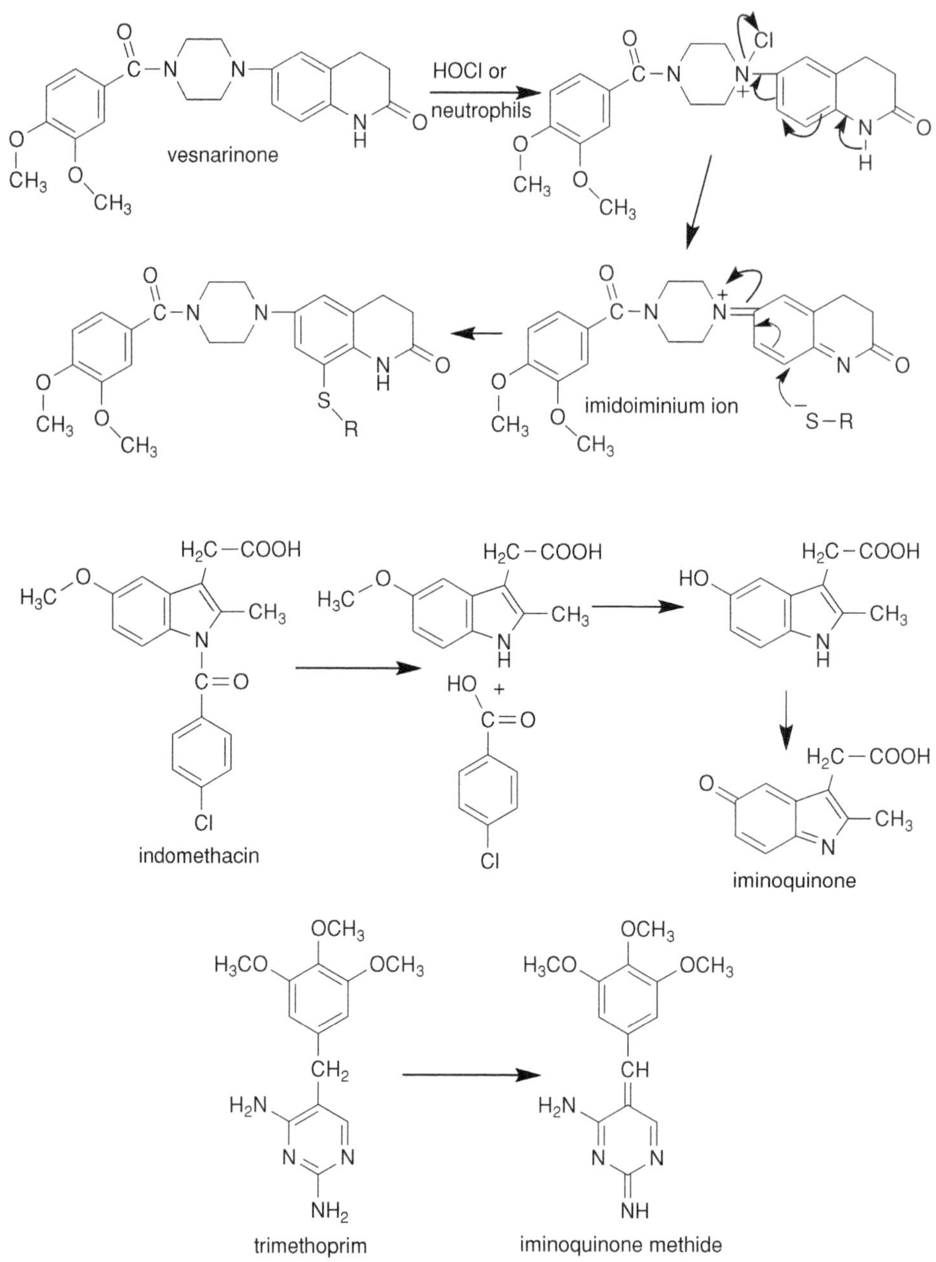

FIGURE 8.15 Examples of quinone-type reactive metabolites that can be viewed as Michael acceptors.

transport reactive isocyanates to sites distant from where the isocyanate was formed and where the reverse reaction can regenerate the isocyanate (28). Isocyanates can also react with amino groups to form a substituted urea.

Isocyanates can be formed by oxidative dehydrogenation (see "Oxidative Dehydrogenation" section and Figure 4.71 in Chapter 4). Isocyanates can also be formed from the oxidation of sulfonylureas (e.g., tolbutamide) (29) and thiazolidinediones (e.g., troglitazone) (30), as shown in Figure 8.19. Both of these classes of drugs are used to treat

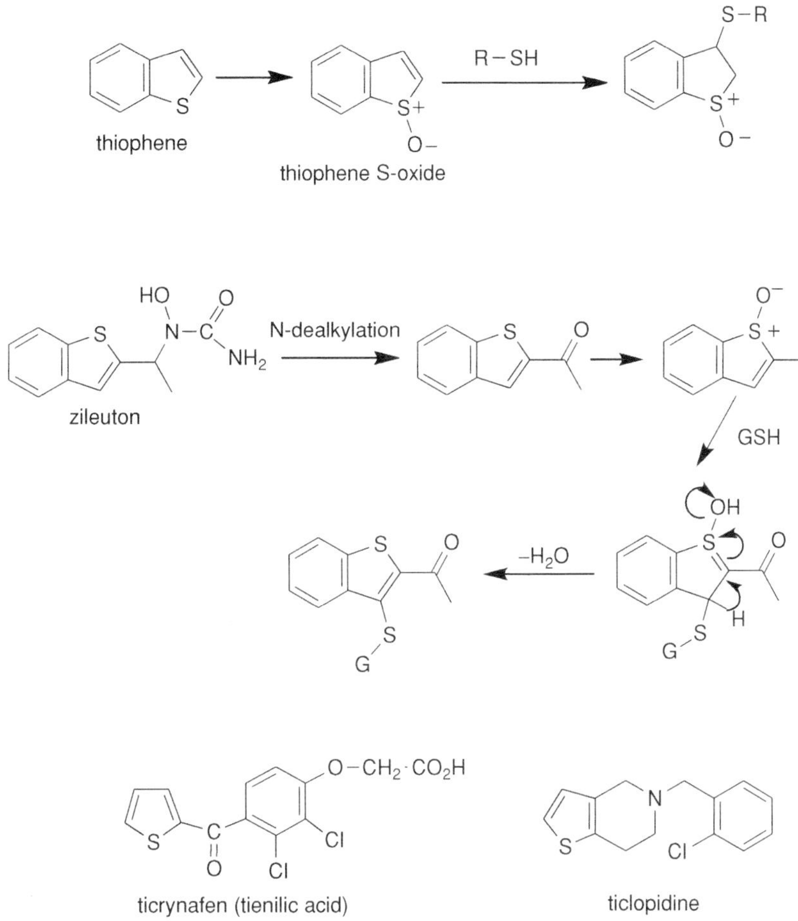

FIGURE 8.16 Thiophenes have the potential to be oxidized to S-oxides that are analogous to acrolein.

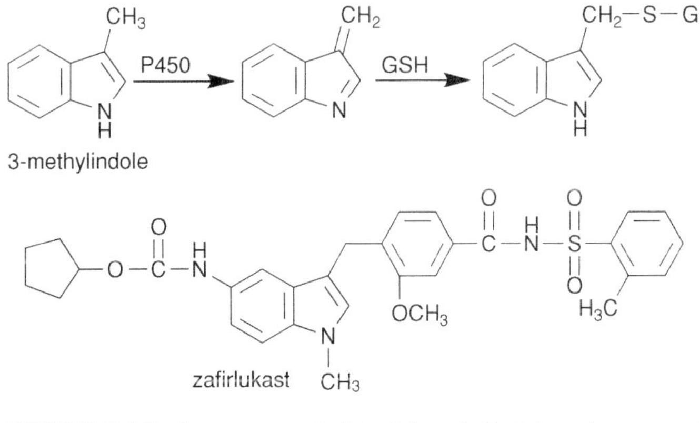

FIGURE 8.17 Reactive metabolite of 3-methylindole and an example of a drug which can form a related reactive metabolite.

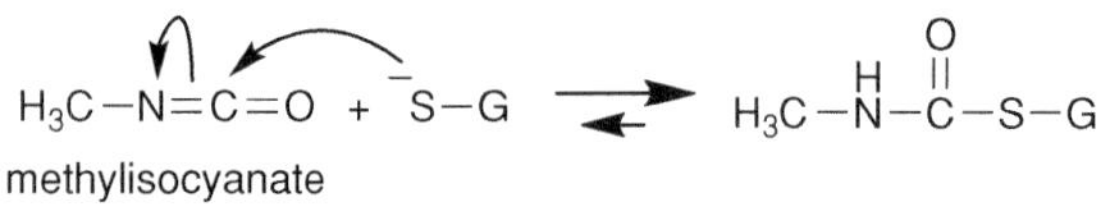

FIGURE 8.18 Reaction of methylisocyanate with a thiol.

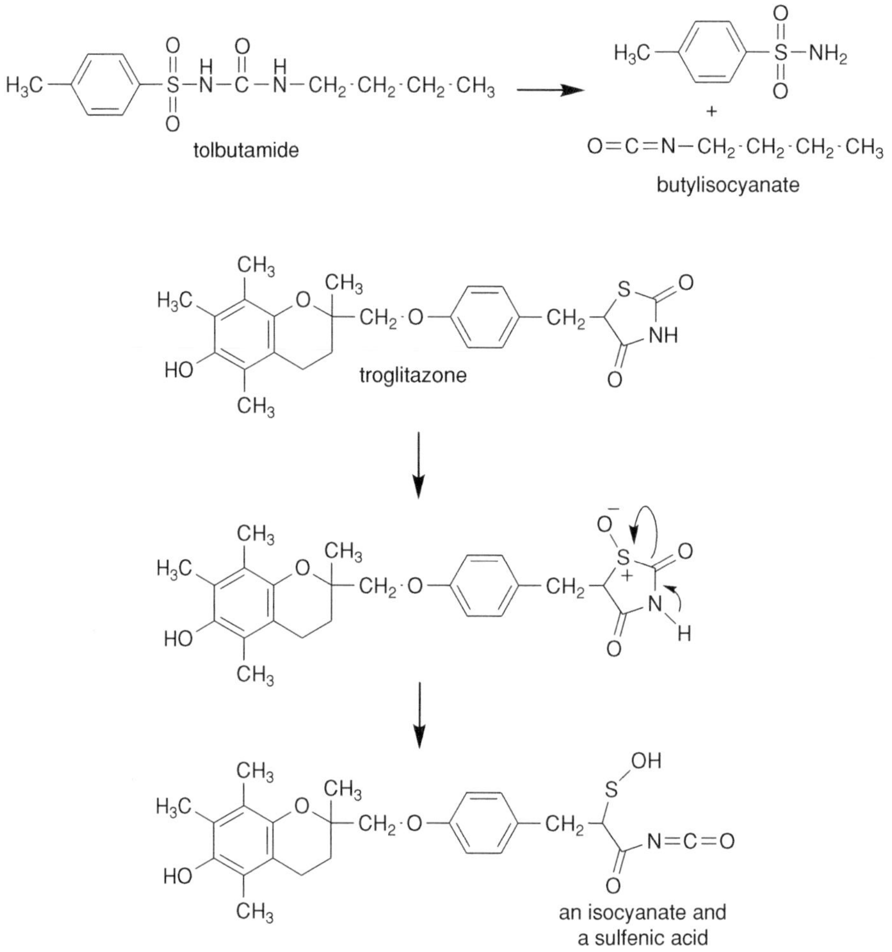

FIGURE 8.19 Metabolism of tolbutamide and troglitazone to isocyanates.

diabetes, and troglitazone—the first in the class thiazolidinedione—had to be withdrawn from the market because of liver toxicity. The oxidation of a thiazolidinedione also results in the formation of a reactive sulfenic acid.

CARBENES

A relatively unique type of reactive metabolite is carbene, i.e., a divalent carbon, which is a proposed intermediate in the oxidation of methylene dioxy-containing compounds. A methylenedioxy group in aromatic compounds is subject to O-dealkylation, e.g., 3,4-methylenedioxyamphetamine, as shown in Figure 8.20. The process generates formic acid and the catechol metabolite as final products. However, in the course of the reaction, a

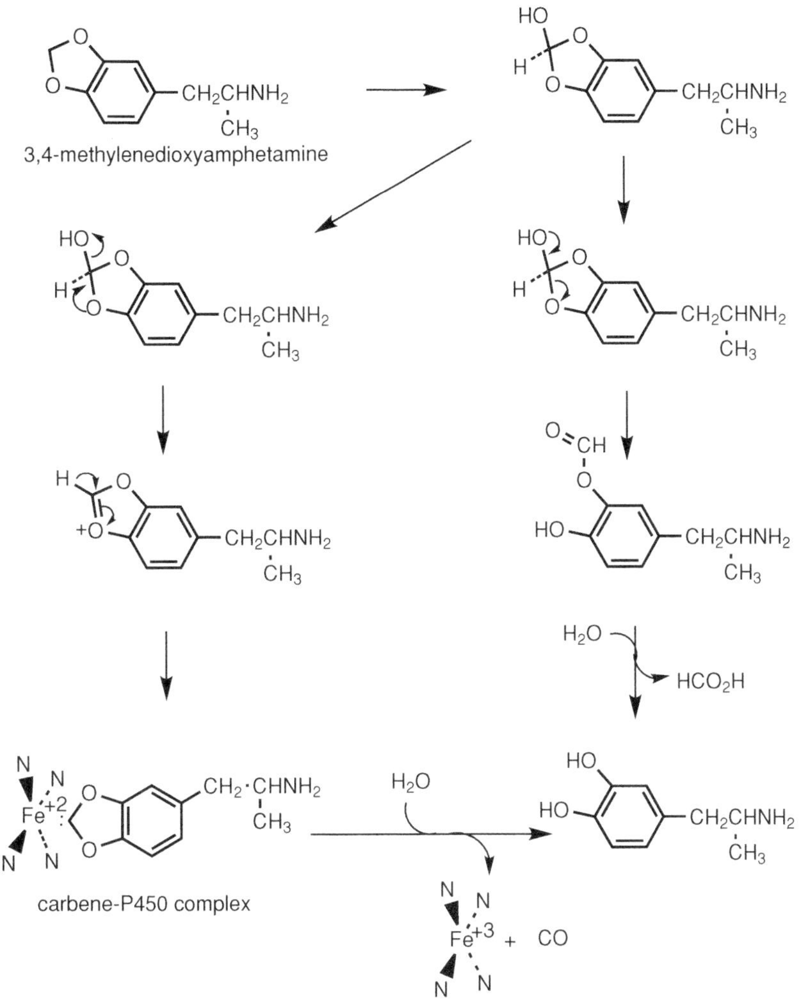

FIGURE 8.20 O-dealkylation of a methylenedioxy-containing drug with formation of a carbene–P450 complex.

portion of the enzyme can be inactivated by formation of what has been termed a metabolic-intermediate complex (31) that is characterized by an absorption peak maximum at 455 nm in the difference spectrum of reduced cytochrome P450. The complexing species that is generated from the methylenedioxy substrate is believed to be a carbene that associates with the iron atom of the ferrous form of the enzyme in much the same way that carbon monoxide does (32). The complex is of moderate stability and thus a quasi-irreversible inhibitor of the enzyme that ultimately dissociates to generate ferric P450, carbon monoxide and the catechol (32). Formation of the complex can be rationalized as arising from a competing pathway in the breakdown of the hydroxylated intermediate to generate the catechol metabolite (33).

Methylenedioxy compounds such as piperonyl butoxide are used to make insecticides more effective by inhibiting the insect enzymes that inactivate the insecticide (Fig. 8.21). Some drugs such as paroxetine and 3,4-methylenedioxymethamphetamine (ecstasy) also contain this functional group.

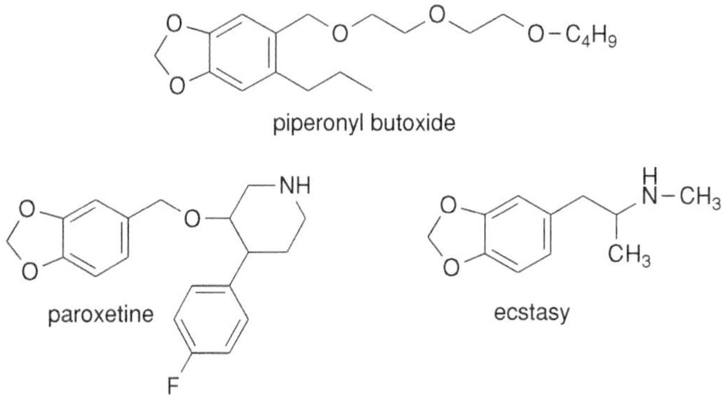

piperonyl butoxide

paroxetine

ecstasy

FIGURE 8.21 Structures of other xenobiotics that contain a methylenedioxy group.

FREE RADICALS

Free radicals are compounds characterized by having an unpaired electron. They are in general highly reactive species, particularly those involving elements of the second row of the periodic table like O, N, C, etc. by the drive to form another chemical bond and complete the valence shell. Because a normal chemical bond consists of two electrons and two electrons only, radicals cannot react covalently with nucleophiles or two-electron species; they can only react with other radicals or they can abstract a hydrogen atom from a neutral molecule to generate a new radical or abstract an electron to form an anion and generate a radical cation. While neutral, they can be considered to be electron deficient, but from the perspective that the addition of another electron to form the anion or the reaction with another radical to form a chemical bond satisfies the rule of eight and completes the second quantum level. For example, they can be deactivated by abstracting an electron from vitamin E or C generating the less reactive vitamin E and C free radicals, or they can abstract a hydrogen atom from other molecules such as glutathione or unsaturated lipids; the latter is shown in Figure 8.22. Abstraction of a hydrogen atom from polyunsaturated lipids is facilitated by the delocalization inherent to the resulting lipid-free radical, which can undergo rearrangement to increase the degree of conjugation as shown. Carbon radicals are very reactive with molecular oxygen, which is a diradical. This leads to a peroxy radical, which in turn abstracts another hydrogen to form a hydroperoxide. The hydroperoxide can also generate more free radicals leading to a chain reaction. In vivo, such reactions are limited by antioxidants such as vitamin E.

One interesting example of the formation of a carbon-centered free radical is the one-electron oxidation of cyclopropylamines as shown in Figure 8.23. Because of the ring strain, the formation of a nitrogen-centered free radical next to a cyclopropylamine leads to opening of the ring and the production of a carbon-centered free radical, which is more reactive than nitrogen-centered free radicals (34). Ultimately, hydrolysis of the iminium ion that is formed leads to loss of the cyclopropyl ring from the molecule.

Although covalent binding of free radicals with other molecules is less common than with electrophiles, they can add to double bonds, most commonly in lipids as shown in Figure 8.24, and this can also lead to chain reactions analogous to hydrogen abstraction reactions. Such oxidations can also lead to the oxidation of protein sulfhydryl groups thus leading to changes in protein structure.

Some metabolites, most commonly aromatic hydroxylamines and quinone-type metabolites, can undergo redox cycling and generate reactive oxygen species, especially

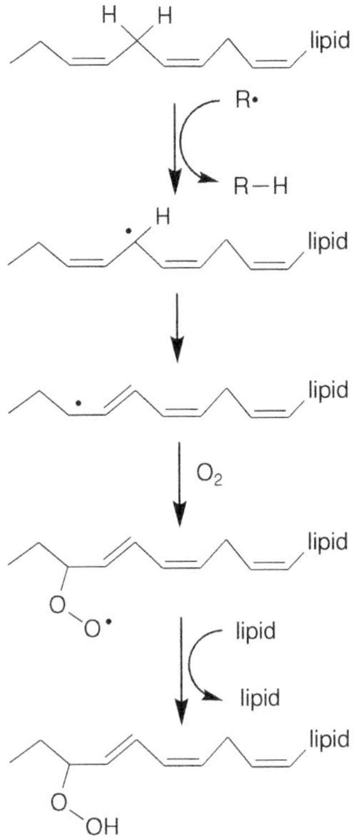

FIGURE 8.22 Reaction of a free radical with an unsaturated lipid and the subsequent rearrangement of double bonds and reaction with molecular oxygen.

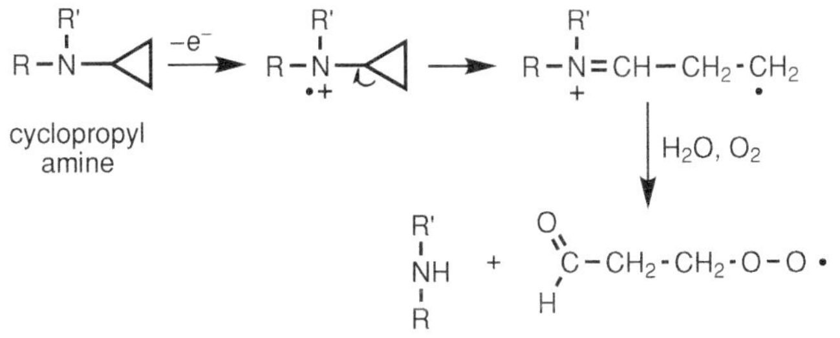

FIGURE 8.23 One electron oxidation of a cyclopropyl amine leading to ring opening and the formation of a carbon-centered free radical and an iminium ion.

superoxide anion. The best example of a molecule that can undergo redox cycling is the herbicide paraquat as shown in Figure 8.25. Paraquat is electron deficient because of the two positive charges and is easily reduced by agents such as vitamin C to a relatively stable free radical. It can, in turn, reduce molecular oxygen to superoxide anion with regeneration of the dication. Thus, in this case, vitamin C can actually increase toxicity by increasing the rate of paraquat reduction. An analogous molecule is formed by the oxidation of 1-methyl-4-phenyl-tetrahydropyridine (MPTP), an impurity formed during the synthesis of a demerol-like narcotic, to MPP$^+$ as shown in Figure 8.25 (35). Administration of MPP$^+$ does not cause toxicity because it cannot get past the blood–brain barrier, but the first oxidation product can enter the brain. It is oxidized by monoamine oxidase B in the brain, and the positively charged MPP$^+$ is actively taken up by dopaminergic neurons. This leads

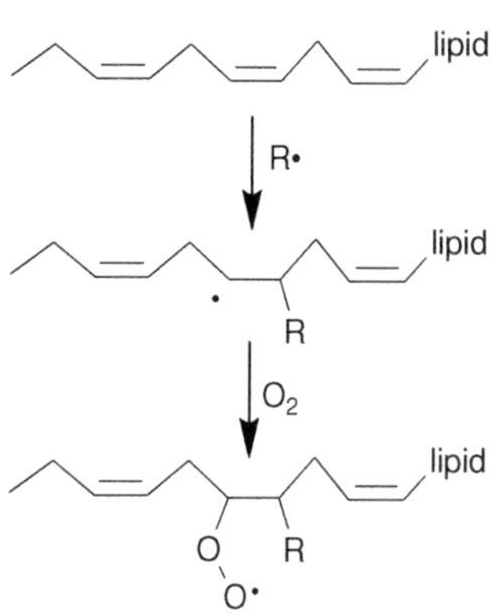

FIGURE 8.24 Addition of a free radical to a lipid double bond resulting in a lipid-free radical.

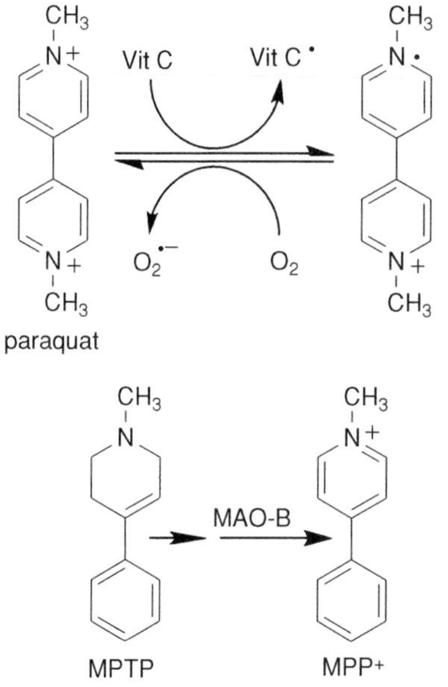

FIGURE 8.25 Redox cycle of paraquat leading to the production of superoxide anion and a similar structure formed by oxidation of MPTP.

to the death of these neurons and results in a Parkinsonian syndrome. It has been proposed that idiopathic Parkinson's disease may also be caused by similar chemicals or drugs that might form similar types of structures.

PREDICTION OF REACTIVE METABOLITE FORMATION

As mentioned earlier, reactive metabolite formation can lead to various types of toxicity. Thus it is important to try to predict which drugs or drug candidates are likely to form reactive metabolites. Several functional groups such as aryl amines/aryl nitro groups, thiophenes, furans, 3-methylindoles, etc. are considered "structural alerts " and are often avoided when synthesizing drug candidates. However, not all drugs containing these functional groups are associated with significant toxicity. Furthermore, there are many pathways that can lead to a reactive metabolite. Therefore, it is difficult to predict all potential reactive metabolites and almost all drugs have the potential to form a reactive metabolite. One strategy that has been used by the pharmaceutical industry is to screen drug candidates for the production of a reactive metabolite in early metabolic studies by searching for glutathione conjugates. However, not all reactive metabolites form glutathione conjugates and some glutathione conjugates are transported into bile and can be further metabolized by gut bacteria so that they are not detected. In addition, conjugation with glutathione usually leads to an unreactive molecule so that if this pathway is very efficient the drug is unlikely to cause toxicity, at least due to a reactive metabolite.

Another strategy is to use radiolabeled analogs to detect irreversible binding to protein (36). It is impossible to do such studies in humans, and it requires a large amount of radiolabeled drug to do whole animal studies. In vitro studies are easier to perform, but they may produce misleading results because the enzyme responsible for bioactivation and/or detoxifying systems may be absent. For example, the first step in the bioactivation of felbamate described earlier in the chapter is the hydrolysis of a carbamate, but this does not occur in the liver (T. Macdonald, personal communication), and therefore if hepatic microsomes or even hepatocytes were used for the study, bioactivation would not be detected. Extrapolation from animals to humans can also lead to misleading results. Again using felbamate as an example, rats and other species that have been studied form significantly less reactive metabolite than humans and therefore studies in animals would underestimate the risk to humans (15).

Although none of the methods for screening drug candidates for the formation of reactive metabolites is perfect, they can add valuable information. One of the most important aspects of drug evaluation is to examine the observed metabolic pathways and to use judgment to evaluate how these metabolites are formed and whether reactive intermediates are likely to be involved. Even if a drug candidate is found to form a large amount of reactive metabolite, it may still not cause an unacceptable incidence of adverse drug reactions— but it must be seen as a major liability. Although it seems likely that screening drug candidates for the formation of reactive metabolites will lead to safer drugs, this has yet to be demonstrated. The usual therapeutic dose is also an important factor that can limit the amount of reactive metabolite formed and drugs given at a dose of 10 mg/day or less are rarely associated with idiosyncratic drug reactions.

REFERENCES

1. Marchand DH, Remmel RP, Abdel-Monem MM. Biliary excretion of a glutathione conjugate of busulfan and 1,4-diiodobutane in the rat. Drug Metab Dispos 1988;16(1):85–92.
2. Lai CC, Miller JA, Miller EC, et al. N-sulfooxy-2-aminofluorene is the major ultimate electrophilic and carcinogenic metabolite of N-hydroxy-2-acetylaminofluorene in the livers of infant male C57BL/6J x C3H/HeJ F1 (B6C3F1) mice. Carcinogenesis 1985;6(7):1037–1045.
3. Miller JA, Miller EC. The metabolic activation and nucleic acid adducts of naturally-occurring carcinogens: recent results with ethyl carbamate and the spice flavors safrole and estragole. Br J Cancer 1983;48(1):1–15.
4. Cribb AE, Nuss CE, Alberts DW, et al. Covalent binding of sulfamethoxazole reactive metabolites to human and rat liver subcellular fractions assessed by immunochemical detection. Chem Res Toxicol 1996;9(2):500–507.
5. Uetrecht J. N-oxidation of drugs associated with idiosyncratic drug reactions. Drug Metab Rev 2002;34(3):651–665.
6. Liu ZC, Uetrecht JP. Clozapine is oxidized by activated human neutrophils to a reactive nitrenium ion that irreversibly binds to the cells. J Pharmacol Exp Ther 1995;275(3):1476–1483.

7. Mathews JM, Bend JR. N-alkylaminobenzotriazoles as isozyme-selective suicide inhibitors of rabbit pulmonary microsomal cytochrome P-450. Mol Pharmacol 1986;30(1):25–32.

8. Stevens GJ, Hitchcock K, Wang YK, et al. In Vitro metabolism of N-(5-Chloro-2-methylphenyl)-N-(2-methylpropyl)thiourea: species comparison and identification of a novel thiocarbamide-glutathione adduct. Chem Res Toxicol 1997;10:733–741.

9. Spahn-Langguth H, Benet LZ. Acyl glucuronides revisited: is the glucuronidation process a toxification as well as a detoxification mechanism. Drug Metab Rev 1992;24:5–47.

10. Boelsterli UA. Xenobiotic acyl glucuronides and acyl CoA thioesters as protein-reactive metabolites with the potential to cause idiosyncratic drug reactions. Curr Drug Metab 2002;3(4):439–450.

11. Anders MW. Glutathione-dependent bioactivation of haloalkanes and haloalkenes. Drug Metab Rev 2004;36(3–4):583–594.

12. Lindstrom AB, Yeowell-O'Connell K, Waidyanatha S, et al. Measurement of benzene oxide in the blood of rats following administration of benzene. Carcinogenesis 1997;18(8):1637–1641.

13. Thakker DR, Yagi H, Levin W, et al. Polycyclic aromatic hydrocarbons: metabolic activation to ultimate carcinogens. In: Anders MW, ed. Bioactivation of Foreign Compounds. Orlando, FL: Academic Press; 1985:177–242.

14. Kitteringham NR, Christie G, Coleman JW, et al. Drug-protein conjugates XII: a study of the disposition, irreversible binding and immunogenicity of penicillin in the rat. Biochem Pharmacol 1987;36:601–608.

15. Dieckhaus C, Miller T, Sofia RD, et al. A mechanistic approach to understanding the species differences in felbamate bioactivation: relevance to drug-induced idiosyncratic reactions. Chem Res Toxicol 2000;28(7):814–822.

16. Iverson SL, Uetrecht JP. Identification of a reactive metabolite of terbinafine: insights into terbinafine-induced hepatotoxicity. Chem Res Toxicol 2001;14(2):175–181.

17. Walsh JS, Reese MJ, Thurmond LM. The metabolic activation of abacavir by human liver cytosol and expressed human alcohol dehydrogenase isozymes. Chem Biol Interact 2002;142(1–2):135–154.

18. Chen LJ, Hecht SS, Peterson LA. Identification of cis-2-butene-1,4-dial as a microsomal metabolite of furan. Chem Res Toxicol 1995;8(7):903–906.

19. Ju C, Uetrecht JP. Oxidation of a metabolite of indomethacin (Desmethyldeschlorobenzoylindomethacin) to reactive intermediates by activated neutrophils, hypochlorous acid, and the myeloperoxidase system. Drug Metab Dispos 1998;26(7):676–680.

20. Uetrecht JP, Zahid N, Whitfield D. Metabolism of vesnarinone by activated neutrophils; implications for vesnarinone-induced agranulocytosis. J Pharmacol Exp Ther 1994;270(3):865–872.

21. Lai WG, Zahid N, Uetrecht JP. Metabolism of trimethoprim to a reactive iminoquinone methide by activated human neutrophils and hepatic microsomes. J Pharmacol Exp Ther 1999;291(1):292–299.

22. Mansuy D, Valadon P, Erdelmeier I, et al. Thiophene S-oxides as new reactive metabolites: formation by cytochrome P450 dependent oxidation and reaction with nucleophiles. J Am Chem Soc 1991;113(20):7825–7826.

23. Joshi EM, Heasley BH, Chordia MD, et al. In vitro metabolism of 2-acetylbenzothiophene: relevance to zileuton hepatotoxicity. Chem Res Toxicol 2004;17(2):137–143.

24. Koenigs LL, Peter RM, Hunter AP, et al. Electrospray ionization mass spectrometric analysis of intact cytochrome P450: identification of tienilic acid adducts to P450 2C9. Biochemistry 1999;38(8):2312–2319.

25. Liu ZC, Uetrecht JP. Metabolism of ticlopidine by activated neutrophils: implications for ticlopidine-induced agranulocytosis. Drug Metab Dispos 2000;28(7):726–730.

26. Nocerini MR, Carlson JR, Yost GS. Electrophilic metabolites of 3-methylindole as toxic intermediates in pulmonary oedema. Xenobiotica 1984;14(7):561–564.

27. Kassahun K, Skordos K, McIntosh I, et al. Zafirlukast metabolism by cytochrome P450 3A4 produces an electrophilic alpha,beta-unsaturated iminium species that results in the selective mechanism-based inactivation of the enzyme. Chem Res Toxicol 2005;18(9):1427–1437.

28. Slatter JG, Rashed MS, Pearson PG, et al. Biotransformation of methyl isocyanate in the rat. Evidence for glutathione conjugation as a major pathway of metabolism and implications for isocyanate-mediated toxicities. Chem Res Toxicol 1991;4(2):157–161.

29. Guan X, Davis MR, Tang C, et al. Identification of S-(n-butylcarbamoyl)glutathione, a reactive carbamoylating metabolite of tolbutamide in the rat, and evaluation of its inhibitory effects on glutathione reductase in vitro. Chem Res Toxicol 1999;12(12):1138–1143.

30. Kassahun K, Pearson PG, Tang W, et al. Studies on the metabolism of troglitazone to reactive intermediates in vitro and in vivo. Evidence for novel biotransformation pathways involving quinone methide formation and thiazolidinedione ring scission. Chem Res Toxicol 2001;14(1):62–70.

31. Franklin MR. The enzymic formation of methylenedioxyphenyl derivative exhibiting an isocyanide-like spectrum with reduced cytochrome P-450 in hepatic microsomes. Xenobiotica 1971;1(6):581–591.

32. Ortiz de Montellano PR, Reich NO. Inhibition of cytochrome P-450 enzymes. In: Ortiz de Montellano PR, ed. Cytochrome P-450. 1st ed. New York: Plenum; 1986.

33. Correia MA, Ortiz de Montellao PR. Inhibition of cytochrome P450 enzymes. In: Ortiz de Montellano PR, ed. Cytochrome P450: Structure, Mechanism, and Biochemistry. 3rd ed. New York: Kluwer/Plenum; 2005:247–322.

34. Shaffer CL, Morton MD, Hanzlik RP. N-dealkylation of an N-cyclopropylamine by horseradish peroxidase. Fate of the cyclopropyl group. J Am Chem Soc 2001;123(35):8502–8508.

35. Salach JI, Singer TP, Castagnoli N, Jr, et al. Oxidation of the neurotoxic amine 1-methyl-4-phenyl-1,2,3,6-tetrahydropyridine (MPTP) by monoamine oxidases A and B and suicide inactivation of the enzymes by MPTP. Biochem Biophys Res Commun 1984;125(2):831–835.

36. Evans DC, Watt AP, Nicoll-Griffith DA, et al. Drug-protein adducts: an industry perspective on minimizing the potential for drug bioactivation in drug discovery and development. Chem Res Toxicol 2004;17(1):3–16.

Practice Problems

1. Work out all of the possible metabolites of lidocaine. Start at the top of the molecule and work toward the other end of it. Just work with one position at a time and do not worry about all of the products that involve combinations of metabolism at two different positions; there are plenty of metabolites without considering such combinations.

lidocaine

2. Provide a sequence of metabolic steps that would lead to the observed metabolite.

a.

b.

c.

d.

167

3. Explain with metabolic sequences why is the ratio of products is same when the two isomers of deuteronaphthalene are oxidized.

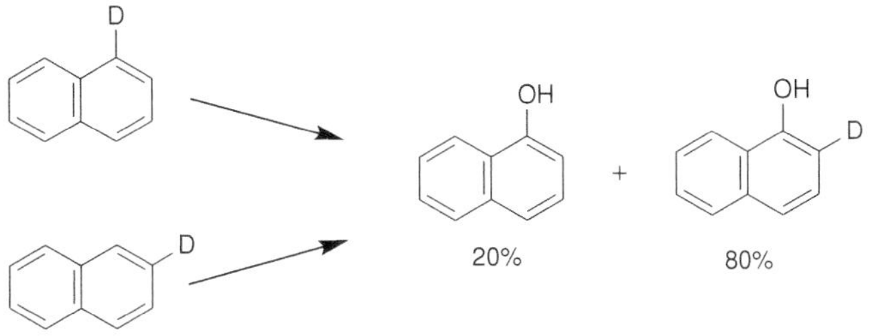

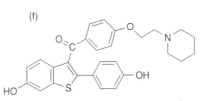

20%   80%

4. Predict whether the following drugs/xenobiotics form glutathione conjugates and, if a conjugate is likely, draw its structure.

(a)

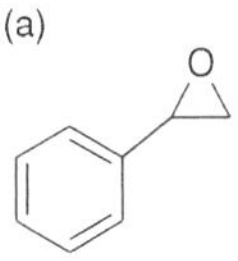

styrene oxide

(b)

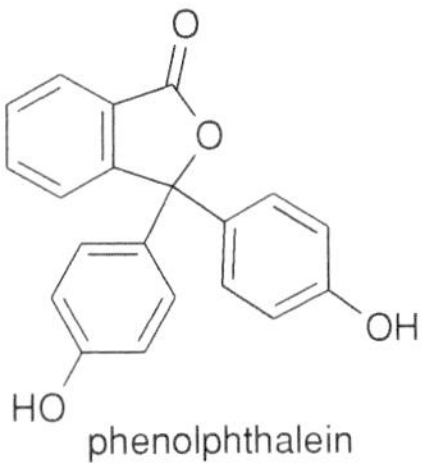

phenolphthalein

(c) OCH₂CO₂H

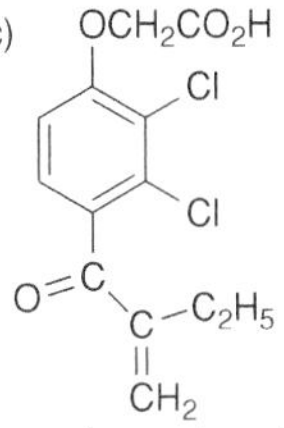

ethacrynic acid

(d)

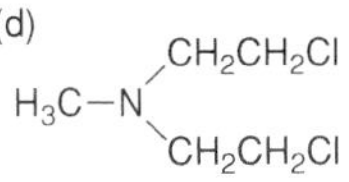

$$H_3C-N\begin{smallmatrix}CH_2CH_2Cl\\CH_2CH_2Cl\end{smallmatrix}$$

mechlorethamine

(e)

$$^-O-S-O-(CH_2)_{11}CH_3$$

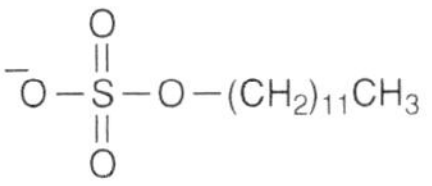

dodecylsulfate

5. Draw the structures of the most likely reactive metabolites of the following drugs/xenobiotics.

(a)

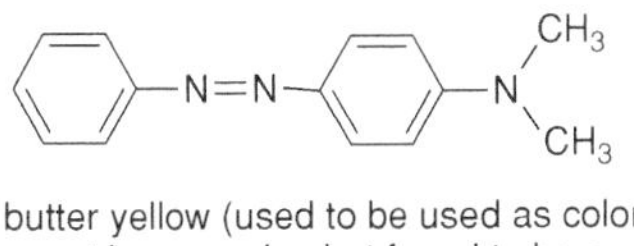

butter yellow (used to be used as coloring agent in margarine but found to be a carcinogen)

(b)

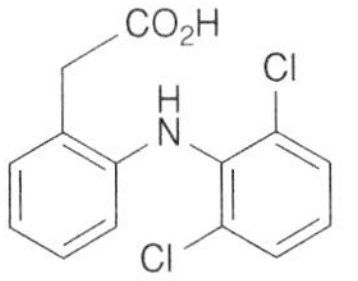

diclofenac (causes idiosyncratic liver toxicity)

(c)

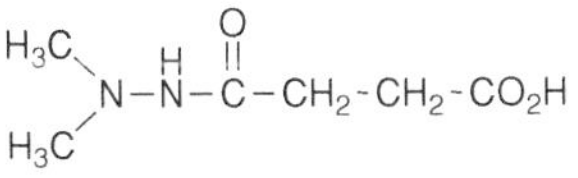

Alar (used as a spray on apples but concern about carcinogenicity)

(d)

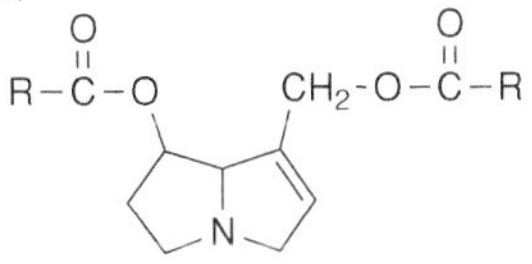

pyrrolizidine alkaloids (carcinogen found in some herbal teas)

(e)

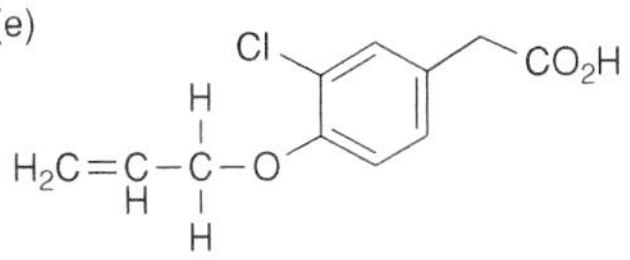

alclofenac (an analgesic withdrawn because of "hypersensitivity" reactions)

(f)

raloxifene (a selective estrogen antagonist that forms several glutathione conjugates)

Answers to Practice Problems

1. Most of the possible metabolites of lidocaine:
    (a) Starting at the top of the molecule, oxidation can lead to an arene oxide (1) or a phenol (4); direct oxidation to the phenol is more likely than involvement of an arene oxide intermediate. The arene oxide can react with glutathione to form the conjugate (2) that can be rearomatize by dehydration to form the product (3). Phenols such as (4) can be conjugated with glucuronic acid and sulfate. The phenol can also be oxidized to the imidoquinone (5), which would be reactive and form a glutathione conjugate (6). This glutathione conjugate can be converted through a series of steps (loss of glutamate followed by loss of glycine to form the cysteine conjugate, which is then acetylated) to form the N-acetylcysteine conjugate (7), otherwise known as a mercapturic acid. Other glutathione conjugates can also undergo the same conversion to mercapturic acids, but this is not shown. The arene oxide (1) can also undergo hydrolysis to form the dihydrodiol (8). The dihydrodiol can be oxidized to form the catechol (9), which can also be formed by further oxidation of the phenol (4). Catechols can undergo conjugation: glucuronidation, sulfation, or methylation. Lidocaine can also form the 2,3 arene oxide (10), which can form a glutathione conjugate. This arene oxide can also undergo an NIH shift to form the phenol (11), which can undergo conjugation: glucuronidation or sulfation. Moving further down the aromatic ring, oxidation of the methyl group leads to a benzylic alcohol (12), which can undergo conjugation: glucuronidation or sulfation. The benzylic alcohol can undergo further oxidation to an aldehyde (13) and further to the benzoic acid (14). The benzoic acid can undergo conjugation to form a glucuronide, a glycine conjugate, and could form a Co-A ester. By symmetry, oxidation of the other methyl group leads to the same products.

Of these pathways, the major one leads to the phenol (4).

(b) Continuing with possible metabolic pathways further down the molecule, the amide nitrogen could be oxidized to a hydroxamic acid (15). The hydroxamic acid can undergo conjugation: acetylation, glucuronidation, and sulfation. The sulfate may be sufficiently reactive to form a glutathione conjugate. The amide bond is hydrolyzed to the aromatic amine (16) and the carboxylic acid (20). The aromatic amine can also undergo conjugation or oxidation to the hydroxylamine (17), which too can undergo conjugation. It can also undergo further oxidation to the nitroso metabolite (18). Nitroso compounds react with glutathione to form a sulfinamide conjugate (19). The carboxylic acid can undergo glucuronidation and may also form a Co-A ester. Oxidation of the adjacent carbon leads to a carbinolamine which spontaneously generates glyoxylic acid (22) and diethylamine (23). Glyoxylic acid is further oxidized to oxalic acid (26), and this can undergo further oxidation to carbon dioxide. Diethylamine can be oxidized to the hydroxylamine (27) and further to the nitrone (28). Moving further down the molecule, oxidation of the nitrogen leads to an *N*-oxide (24). Oxidation of the ethyl carbon next to the nitrogen leads to a carbinolamine (25) that spontaneously leads to the loss of the ethyl group as acetaldehyde, which would be further oxidized to acetic acid. The secondary amine (29) can undergo a second N-dealkylation to form a primary amine (31). The primary amine can undergo oxidation to the hydroxylamine (32) and further to the oxime (33). In the intact molecule, oxidation of the carbon that led to N-dealkylation leads to the carbinolamine (34). In addition to loss of acetaldehyde, this carbinolamine can dehydrate to form the iminium ion (35). This iminium ion is electrophilic and can be attacked by the amide nitrogen to form a new ring (36) as described in "N-Dealkylation/Deamination" section in chapter 4. In principle, the end methyl group could also be oxidized to an alcohol and on to an aldehyde and carboxylic acid although this is unlikely to be a significant pathway.

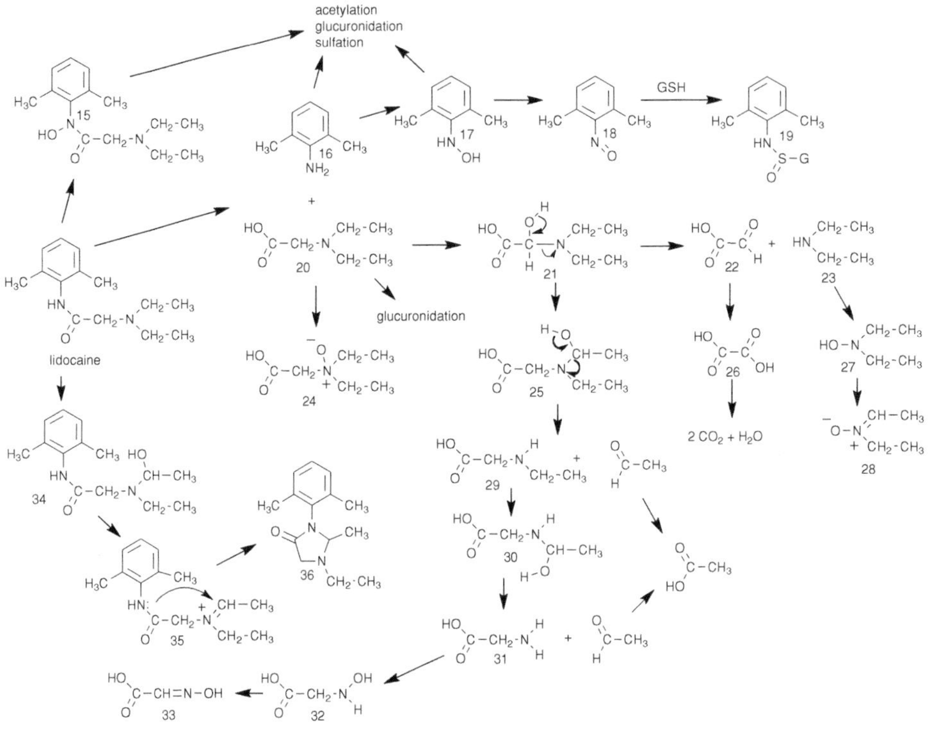

The major metabolites of lidocaine formed from this part of the molecule are the amine (16), which is further oxidized to the para-phenol, N-dealkylation of the parent drug with loss of acetaldehyde to form the secondary amine, and a second N-dealkylation to form the primary amine (structure not shown). Although there are only a few major metabolites of lidocaine, with sensitive analytical methods it is likely that hundreds of minor metabolites could be detected.

2. (a)

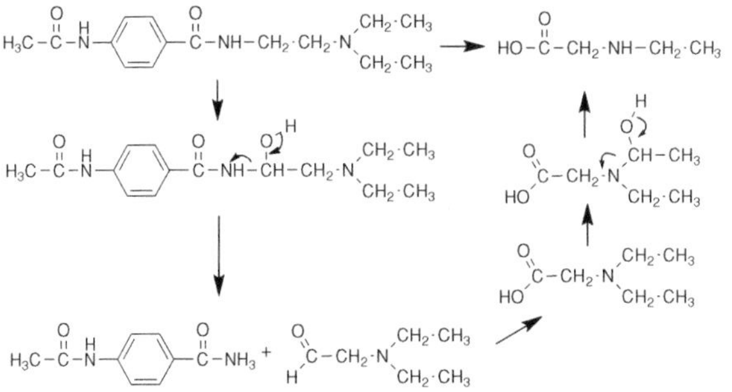

(b)

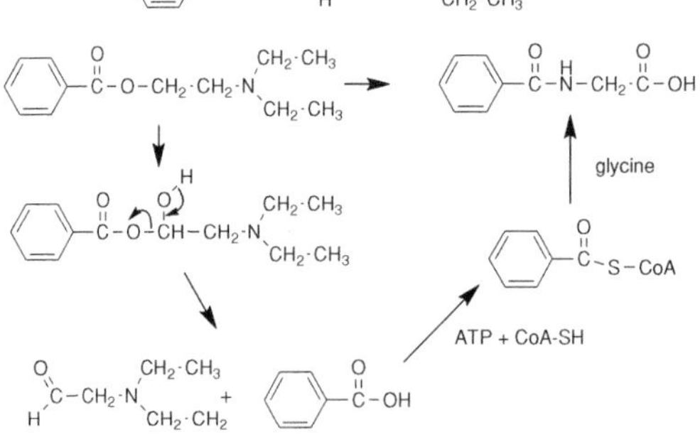

(c)

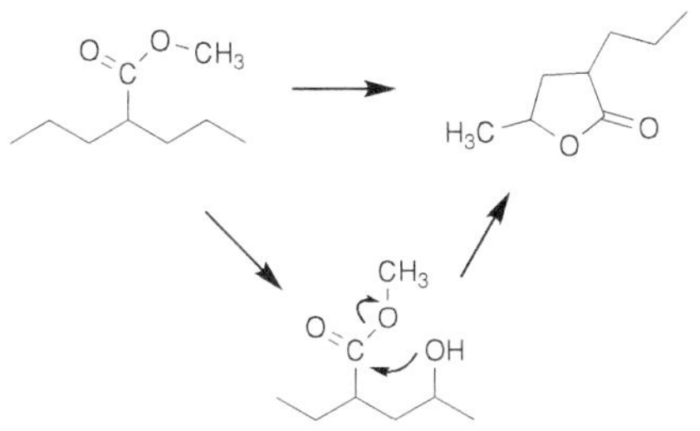

(d)

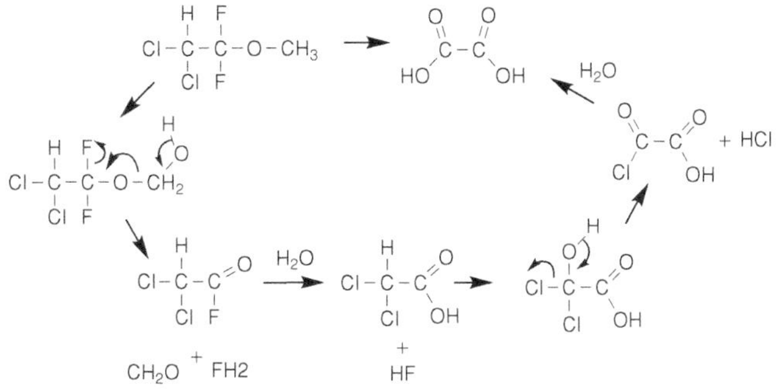

3. They both produce the same intermediate, and because of the deuterium isotope effect, more hydrogen is lost than deuterium.

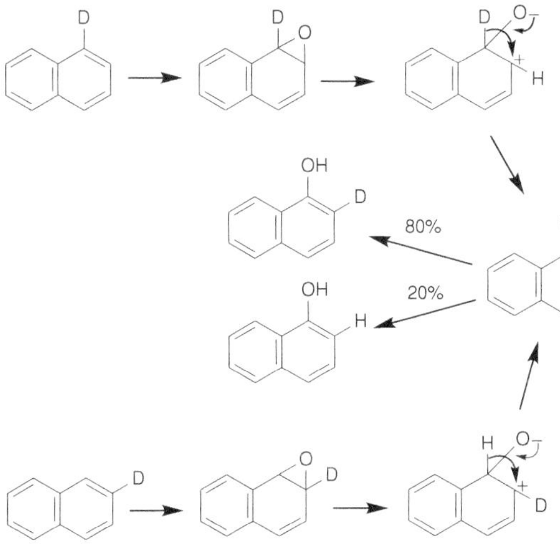

4. (a) Styrene oxide reacts with glutathione, catalyzed by glutathione transferase, to form all of the four possible isomers.

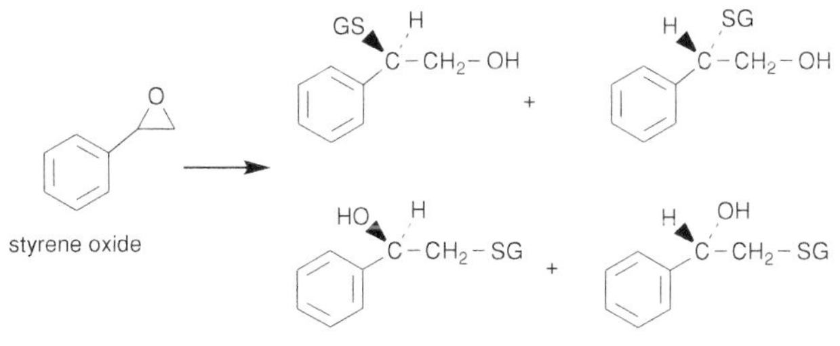

(b) Phenolphthalein is a pH indicator and its alkaline form is a quinone methide that presumably reacts with glutathione. It was previously used as a laxative, but after finding it to be mutagenic it was removed from such products.

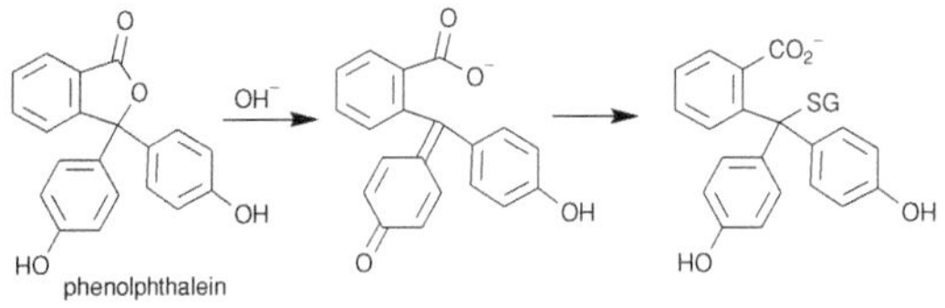

(c) Ethacrynic acid is a Michael acceptor and directly reacts with glutathione to form a conjugate as shown in Figure 16 in chapter 7.

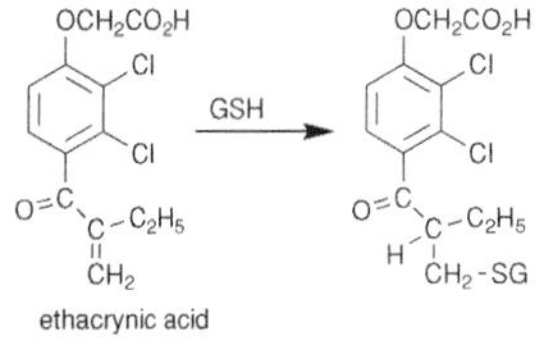

(d) Mechlorethamine is an alkylating agent. Although alkyl chlorides are not usually very reactive, the adjacent nitrogen forms a very reactive aziridium ion as shown in Figure 10 in chapter 8. The first reaction with glutathione is shown, but the process is repeated with the other ethyl chloride so that the molecule can react with two glutathione molecules.

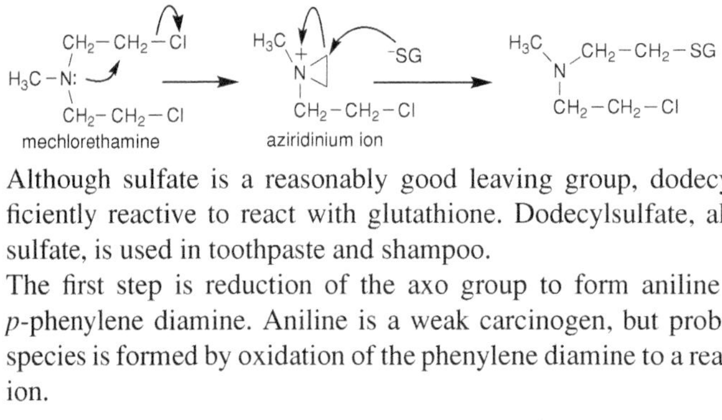

(e) Although sulfate is a reasonably good leaving group, dodecylsulfate is not sufficiently reactive to react with glutathione. Dodecylsulfate, also known as lauryl sulfate, is used in toothpaste and shampoo.

5. (a) The first step is reduction of the axo group to form aniline and N,N-dimethyl-p-phenylene diamine. Aniline is a weak carcinogen, but probably the more toxic species is formed by oxidation of the phenylene diamine to a reactive imine iminium ion.

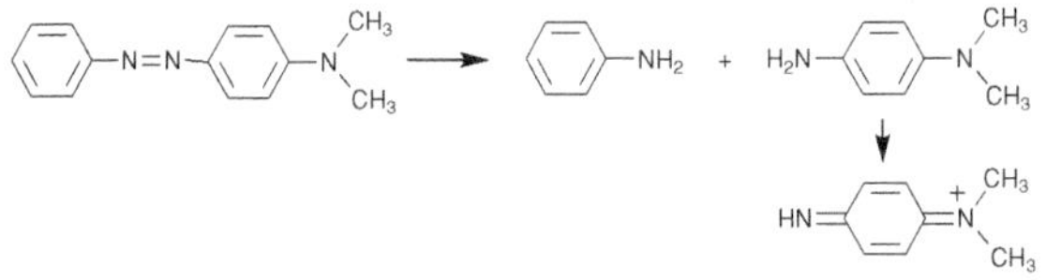

(b) Although diclofenac forms reactive acyl glucuronides and thioesters, both aromatic rings are also oxidized para to the amine and further oxidation leads to reactive iminoquinones.

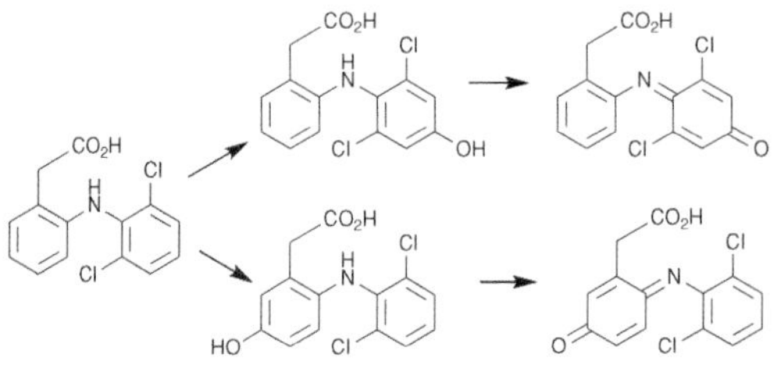

(c) The first step is hydrolysis which results in dimethylhydrazine, a known carcinogen. Presumably this has to be N-alkylated to form methylhydrazine that can be oxidized to the diazine and on to nitrogen gas and a reactive methyl carbocation.

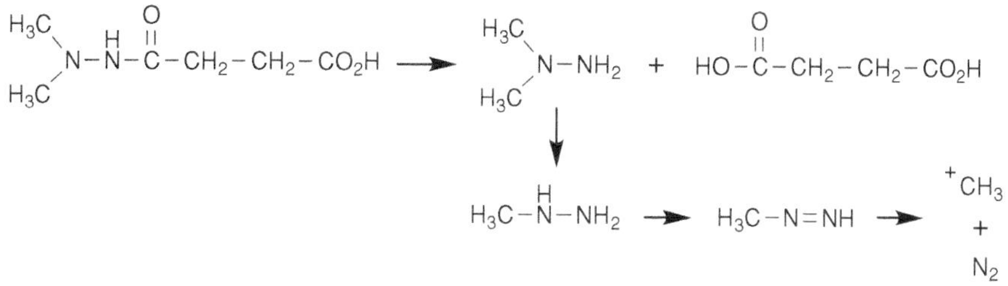

(d) The first step is a dehydrogenation to produce a pyrole ring. This can lose the acyl group to produce a very reactive positively charged methide.

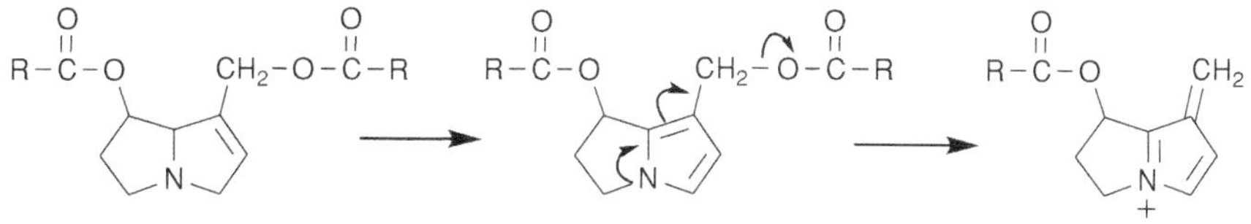

(e) One possible reactive metabolite is an epoxide formed by oxidation of the alkene. However, an even more likely pathway is O-dealkylation which leads acrolein, a known toxic agent. It is interesting that such a structure would ever be developed as a drug.

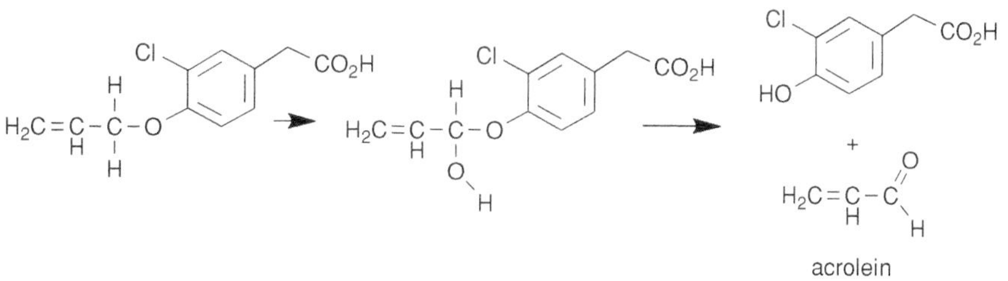

acrolein

(f) Although there are other potential reactive metabolites, the most direct is oxidation of the phenols to a quinone-like structure that spans three ring systems as shown next. The major glutathione conjugates are formed by attacking at the positions marked with arrows.

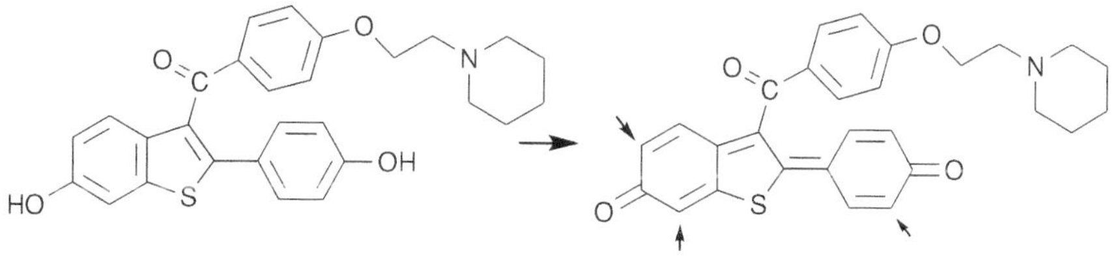

Index

T - #1042 - 101024 - C0 - 254/178/10 - PB - 9781420061031 - Gloss Lamination